MRI Atlas of Human White Matter

MRI Atlas of Human White Matter

SUSUMU MORI
SETSU WAKANA
LIDIA M. NAGAE-POETSCHER
PETER C.M. VAN ZIJL

Johns Hopkins University School of Medicine,
Department of Radiology,
Division of MRI Research,
Baltimore, MD, USA
F.M. Kirby Research Center for Functional Brain Imaging,
Kennedy Krieger Research Institute,
Baltimore, MD, USA

Research Funded by:
National Center for Research Resources
and Human Brain Project,
National Institutes of Health

Collaborator:
Dr. Barbara J. Crain
John Hopkins University School of Medicine
Department of Pathology

2005

ELSEVIER

Amsterdam – Boston – Heidelberg – London – New York – Oxford – Paris –
San Diego – San Francisco – Singapore – Sydney – Tokyo

ELSEVIER B.V.
Radarweg 29
P.O. Box 211, 1000 AE
Amsterdam, The Netherlands

ELSEVIER Inc.
525 B Street, Suite 1900
San Diego, CA 92101-4495
USA

ELSEVIER Ltd
The Boulevard, Langford Lane
Kidlington, Oxford OX5 1GB
UK

ELSEVIER Ltd
84 Theobalds Road
London WC1X 8RR
UK

First edition 2005

Library of Congress Cataloging in Publication Data
A catalog record is available from the Library of Congress.

British Library Cataloguing in Publication Data
A catalogue record is available from the British Library.

ISBN: 0-444-51741-3

⊗ The paper used in this publication meets the requirements of ANSI/NISO Z39.48-1992 (Permanence of Paper).
Printed in Italy.

Contents

Preface

Recent advances in non-invasive imaging techniques have greatly enhanced the knowledge of brain anatomy and its relationship to brain function. Using MRI, detailed structural information can be obtained that allows clear separation of gray matter and white matter structures in the living human brain. This has set the stage for a next level of investigation, that of identification of functional regions and their connections. This *in vivo* localization of areas of functional activity has been and is being actively pursued by a range of different imaging modalities, including single photon emission computed tomography (SPECT), positron emission tomography (PET), magnetoencephalography (MEG), and functional MRI. Until recently, however, it was not possible to identify, *in vivo*, the three-dimensional (3D) anatomy of the major white matter fiber bundles containing the neuronal projections connecting the different functional regions of the brain. These pathways determine the relationship between complex behavior and brain structure and are fundamental to the understanding of large-scale neuro-cognitive networks.

Because of the vast number of neurons (more than 100 billion) and sheer complexity of neuronal connectivity (a single neuron may communicate with a score of neurons via dendrites and an axon with complex branchings), no single technique can even scratch the surface of the brain neural network. One approach to study axonal organization is by chemical tracer techniques, in which a tracer is injected in a group of neurons and actively transported along their axons. Even though this can provide direct evidence of axonal projections, this microscopic technique (which can study only a small number of neurons at a time) is not suitable to study global neuronal anatomy. An alternative, more macroscopic technique is the approach in which a lesion is made and neural degeneration in distal areas is studied. However, chemical tracer and lesion techniques can obviously not be applied to humans. Inadvertent lesions, such as those due to traumatic injuries and stroke, have been important sources of our understanding of human neural architecture, but do not allow a comprehensive mapping of the macroscopic features of normal white matter. Until recently, the most macroscopic level of information of brain axonal architecture (axonal bundles and white matter tracts) has come from postmortem examination. By using special preparation schemes, axonal bundles can be peeled off layer-by-layer and their macroscopic arrangements can be visualized. We can find various sources of drawings and pictures of white matter tracts based on this type of classical anatomical studies. However, this technique can only show the revealed surface of the brain and it is often difficult to understand 3D configurations of tracts from them. Recently, a unique new MRI modality, called diffusion tensor imaging (DTI), has emerged, allowing the three-dimensional study of the large white matter fiber bundles at macroscopic resolution (millimeter scale). The goal of this atlas is to provide three-dimensional *in vivo* illustration of these structures in human brain.

In this atlas, the reconstruction of 15 prominent white matter tracts in the cerebral hemisphere and 5 tracts in the brainstem is presented. These tracts were chosen because their core regions have been well described previously in classical neuroanatomy. Our 3D reconstruction approaches were designed to faithfully reconstruct these core regions by using *a priori* anatomical knowledge (see Chapter 2). In the deep white matter, the white matter bundles tend to be more compacted and more discretely identifiable as a specific tract system. As will be described later, reconstruction of the subcortical white matter tends to be less reliable and tract assignment becomes less robust as each tract starts to diverge into the cortex. For this reason, we avoided assigning tracts from the subcortical regions in this atlas.

One of the purposes of this atlas is education on brain white matter anatomy. Understanding of this anatomy requires an understanding of the 3D configuration of the white matter tracts. This is not an easy task because the white matter tract trajectories are often complex and their mutual relationships are intricate. In order to understand their architectures, one needs to have a flexible and comprehensive 3D visualization of these tracts and a 2D registration with respect to other white matter tracts and gray matter structures. In order to fulfill these requirements, the various white matter tracts depicted in this atlas are assigned different colors and visualized three-dimensionally. Their locations are superimposed on conventional 2D images with three different slice orientations (axial, coronal, and sagittal) and at multiple slice levels to appreciate their precise spatial locations and relationships with other anatomical structures.

Another purpose of the atlas is to provide assignment of anatomical structures on so-called 2D color maps, which display the orientation of the fiber bundles in each brain slice. This is motivated by our belief that color maps will be an important clinical tool in the near future. DTI is becoming more and more available to clinical scanners. Color maps provide intra-white matter anatomy that can not be obtained by conventional MRI. Using this technique, clinicians are now capable of assessing the status of specific white matter tracts in patients with developmental abnormalities, tumors, or neurodegenerative diseases, to name a few. However, for many radiologists, the color map is still a novel imaging modality and training is required to read them. To fully evaluate the color maps, it is essential to understand white matter anatomy three-dimensionally as well as how each tract is revealed in a given 2D observation plane. In this atlas, color maps are presented at multiple slice levels and orientations and the white matter structures are identified, assigned, and annotated by comparison with their reconstructed 3D trajectories.

Acknowledgment

We would like to thank Human Brain Project and National Center for Research Resources, NIH for finantial support. The presented work would not be possible without extraordinary support from Philips Medical System. We also had immense help from Mrs Terri Brawner and Miss Kathleen Kahl for MRI studies. Finally, a special thanks to Mr. Andrew Gent and Mrs Jūratė Murauskienė, Dr Zigmantas Kryžius (both at VTEX, Vilnius) for their help during the preparation and production of this book.

CHAPTER 1

Introduction

This is a human brain atlas based on diffusion tensor imaging (DTI), which is a relatively recent type of MRI. In order to appreciate the images in this atlas it is necessary to understand how they are generated. In this section we therefore provide some general MRI background, followed by the basics of DTI and overview of the principles and limitations of the fiber reconstruction technology.

1. Magnetic Resonance Imaging (MRI)

MR images consist of individual picture elements (pixels) with different intensities (brightness). When evaluating MR images, two important parameters need to be considered: spatial resolution and contrast. In modern MRI, the spatial resolution (pixel size) is on the order of 1–3 mm or even smaller, revealing great detail of brain anatomy. Contrast is created by differences in pixel intensities among different areas of the brain. Conventionally, MR contrast has been mostly based on differences in tissue water relaxation times, such as T_1 and T_2, which can be used to distinguish various brain regions such as the cortex, the deep gray matter, and the white matter. However, conventional MR has not been successful in providing good contrast within the white matter, which usually looks rather homogeneous, no matter how high the spatial resolution is. From an MRI point of view, the white matter generally appears as if it were a fluid-like homogeneous structure, which, of course, is not the case.

The white matter consists of axons connecting different areas of the brain. These axons tend to form "bundles" together with other axons, and, depending on the destinations, the diameters of these bundles (called white matter tracts) can be as large as a few centimeters. Some of them, such as the corpus callosum and the anterior commissure at the mid-sagittal level, are clearly visible in conventional MRI. However, most of these bundles cannot be individually identified by MRI or even in postmortem brain slices. This is because the majority of these bundles have similar chemical compositions and MR signatures (T_1 and T_2 relaxation times). It is very difficult to appreciate where the specific white matter tracts of interest are and how they are spatially related to one another. Actually, conventional MRI is also not able to provide excellent contrast for intra-gray matter structures either. One cannot tell differences in cytoarchitecture among different cortical layers or between cortical areas. Nevertheless, gyral and sulcal patterns can be used as visual clues for cortical anatomy discrimination. Unfortunately, equivalent visual clues are not often available when one tries to identify white matter tracts.

2. Diffusion Tensor Imaging (DTI)

2.1. The diffusion tensor and the diffusion ellipsoid

MRI is an imaging technique that detects proton signals from water molecules. Images thus reflect water density and properties as a function of position in space. In addition, MRI can be used to measure local chemical and physical properties of water. Two of such properties are molecular diffusion and flow. MRI can assess both, but their measurement methods are different and they should not be confused.

The diffusion process is a reflection of thermal Brownian motion. This process can be understood in a simple manner by comparing it to the evolution in the shape of a stain after ink is dropped on a piece of paper. Usually the drop turns into a circle (Gaussian distribution) that grows over time. The faster the diffusion is, the larger the diameter of the circle, and the extent of the diffusion can be estimated from this. Because the extent of the stain is equivalent in all directions, the diffusion is called isotropic. However, if the paper consists of a special fabric which is woven with dense vertical fibers and sparse horizontal fibers, the stain will have an oval shape elongated along the vertical axis. This is called anisotropic diffusion. A similar process happens in the brain where water tends to diffuse preferentially along axonal fibers. If ink were injected inside the brain white matter, the shape of its distribution would be elongated along the axonal tracts. Inside gray matter, the diffusion process is more random because of the lack of aligned fiber structures, and the shape of the ink would be more spherical.

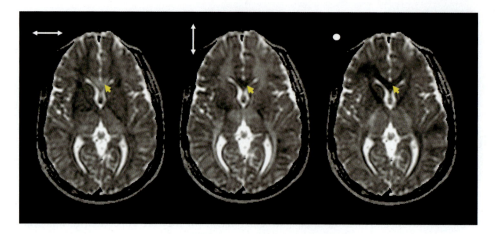

Fig. 1. Diffusion constant maps of the same slice measured along three different orientations (indicated by white arrows and dot). For the third image, the orientation is perpendicular to the plane. In these images, brightness represents the extent of diffusion (magnitude of the diffusion constant). Diffusion in white matter is more anisotropic than that in gray matter. For example, the corpus callosum (yellow arrows) has a high diffusion constant when water diffusion is measured along the horizontal axis (first image) but a low diffusion constant when applying diffusion weighting along the vertical axis (second image) or perpendicular to the plane (third image).

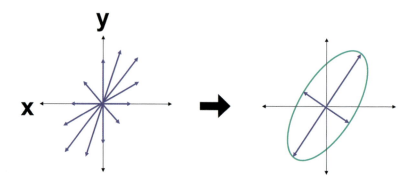

Fig. 2. Using diffusion measurements along multiple axes (left, the length of blue arrows represent the magnitude of diffusion constants), the anisotropic diffusion process can be delineated in detail. In diffusion tensor imaging, the measurement results are fitted to a simple symmetric 3D ellipsoid (or oval in this 2D simplified representation).

One of the unique aspects of diffusion measurements by MRI is that the extent of water diffusion can be measured along one predetermined axis and that this axis can be set arbitrarily. Diffusion measurements can be combined with imaging, from which the effective diffusion constant at each pixel is measured (called apparent diffusion constant map). Examples of such diffusion constant maps are shown in Fig. 1. From these images, it can be clearly seen that water diffusion inside the brain is anisotropic: the extent of the measured diffusion constants depends on the measurement orientation, especially in the white matter.

The extent of anisotropy of the water diffusion can be delineated in more detail by measuring the diffusion constant along multiple axes. Due to the complexity of the underlying cytoarchitecture (in addition to noise), the measurements may yield a very complex pattern of anisotropy. However, if there is a homogeneous cytoarchitecture within a pixel (typically $3 \times 3 \times 3$ mm^3), the magnitude of diffusion (or the shape of the ink stain in our analogy) can be described by a simple symmetric 3D ellipsoid (Fig. 2).

Diffusion tensor imaging is a technique in which diffusion is measured in a series of different spatial directions, from which the shape and orientation of the diffusion ellipsoid is determined at each image pixel by fitting the measurement results to the ellipsoid describing the average local cytoarchitecture. During this measurement and fitting process, we use a 3×3 tensor, from which the name, diffusion tensor imaging, is derived. The fitting process for determining the ellipsoid is a mathematical process called diagonalization of the tensor. The 3×3 diffusion tensor has 9 elements, but there are only 6 independent numbers because it is a symmetric tensor, which makes sense, given the fact that the tensor uniquely determines the diffusion ellipsoid. Six parameters are needed to completely describe the magnitude and orientation of the 3D ellipsoid (Fig. 3) and determination of these 6 parameters is the target of DTI. Thus, determination of the tensor elements requires measurement of diffusion constants along at least 6 spatial directions.

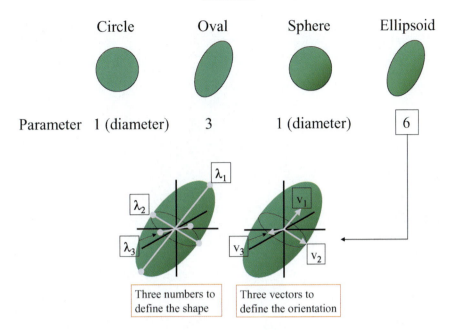

Fig. 3. Parameters required to mathematically describe a circle, oval, sphere, and ellipsoid. The 6 parameters required for an ellipsoid are three eigen-values (λ_1, λ_2, and λ_3) that define the shape of the ellipsoid and three eigenvectors (v_1, v_2, and v_3) that define the orientation of the ellipsoid.

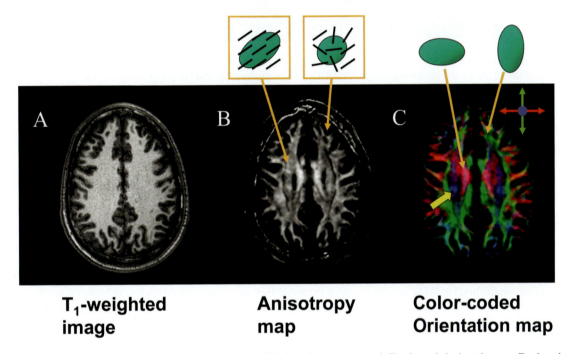

Fig. 4. Examples of (A) conventional MRI (T_1-weighted image), (B) diffusion anisotropy map, and (C) color-coded orientation map. For the color-coded images, the intensity (brightness) is proportional to the anisotropy information (same as in anisotropy map (B)) and the different colors, red, green, and blue represent fibers running along the right-left, anterior-posterior, and inferior-superior axes, respectively. The yellow arrow points at the corona radiata.

2.2. Two-dimensional visualization of DTI results

Once the 6 tensor parameters are determined at each pixel, the information is usually reduced to produce different types of images that can be appreciated visually. It is almost impossible to visualize all 6 parameters in an intuitive way in a single image, and there are two types of presentations that are generally used. One is the anisotropy map and the other is the orientation map or color map. The anisotropy map provides information on the extent of elongation (anisotropy) of the diffusion ellipsoid (Fig. 4B). In this kind of map, the white matter appears brighter (more anisotropic) than the gray matter, showing brain areas with coherent fiber structures. It is believed that such factors as axonal density, myelination, and homogeneity in the axonal orientation affect the degree of diffusion anisotropy. In the color map, the angle of the longest axis of the ellipsoid (v_1 in Fig. 3) is visualized using three principal colors (red (R), green (G), and blue (B)), under the assumption that it indicates the dominant fiber orientation (Fig. 4C). The signal intensity is weighted by the diffusion

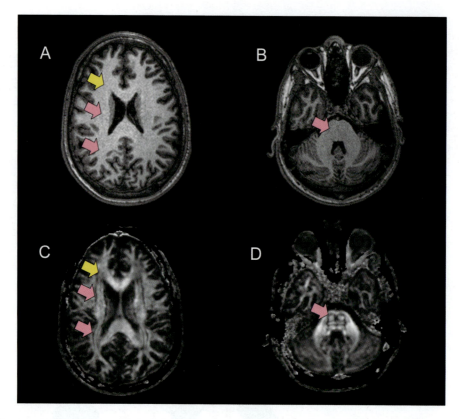

Fig. 5. Typical white matter regions where low anisotropy is observed due to a mixture of axonal tracts bearing different orientations in the same pixel. Images in (A) and (B) are T_1-weighted images and (C) and (D) are anisotropy maps of the corresponding slice locations.

anisotropy. In this kind of map, any arbitrary angle can be represented by a mixture of RGB colors. For example, in Fig. 4, fibers running 45° between the right-left (red) and anterior-posterior (green) axes would be labeled by a yellow (red + green) color.

By comparing these three images, it is clear that the color-coded orientation map carries by far the greatest amount of information regarding white matter anatomy. In the conventional T_1-weighted MR image on the left, the white matter presents as a homogeneous field, whereas many internal structures are visible in the color map. For example, the blue-colored fiber indicated by a yellow arrow in Fig. 4C is the corona radiata, which could not be identified in the T_1-weighted image. Thus, DTI can be used to parcellate the white matter into substructures.

There is one important limitation of DTI that is essential to understand when reading color maps. This is due to the DTI assumption that there is a homogeneous fiber structure within a pixel. Because of the relatively large size of the pixels in DTI data (2–3 mm), a pixel often contains axonal tracts with multiple orientations. Inclusion of tracts with different orientations causes the pixel to loose anisotropy. As a matter of fact, gray matter displays low anisotropy not because there are no axonal fibers in it, but, because the fiber architecture is convoluted with respect to the pixel size. A similar situation can happen in the white matter as well. One of the most notable regions for this is at the junction between the corpus callosum and the corona radiata (yellow arrow in Fig. 5C), causing a dark region to occur in the anisotropy map. These low anisotropy strings in the white matter occur in many other boundaries between prominent fibers of different orientations, as indicated by the pink arrows in Fig. 5 C and D. These areas do not have any representation in T_1-weighted images (Fig. 5 A and B) in which the gray and white matter contrast is provided predominantly by myelin water concentration and relaxation properties. Definitly, these low anisotropy areas should not be interpreted as low contents of axonal tracts.

2.3. Three-dimensional reconstruction of white matter tracts

In chemical tracer techniques, the tracer may travel the entire length of an axonal projection. Using DTI, such a long-range tracing of a single axon is not possible, as one can obtain only pixel-by-pixel information of water diffusion. Many axons may enter a pixel and exit from it. Consequently, the information of a specific trajectory degenerates within this pixel. However, if many axonal tracts form a large bundle (i.e., white matter tract) and the bundle is sufficiently large to contain many pixels across its diameter, a 3D rendering of such a bundle can be reconstructed based on the DTI data simply by connecting the fiber orientation information pixel-by-pixel (for more information see Methods section). An example of a 3D tract reconstruction and comparison with a postmortem specimen is shown in Fig. 6. The strong similarity of the two images suggests that the DTI-based fiber reconstruction technique can reveal the macroscopic anatomy of white matter tracts. On the other hand, it is also true that both DTI images and postmortem dissections cannot trace the trajectory of

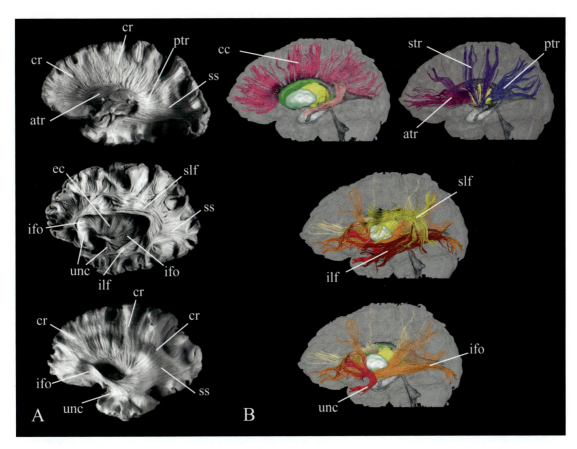

Fig. 6. Comparison between postmortem preparation (A) and DTI-based reconstruction results (B). Abbreviations are atr: anterior thalamic radiation; cc: corpus callosum; cr: corona radiata; ec: external capsule; ifo: inferior fronto-occipital tract; ilf: inferior longitudinal fasciculus; ptr: posterior thalamic radiation; ss: sagittal stratum; slf: superior longitudinal fasciculus; str: superior thalamic radiation; unc: uncinate fasciculus. Copyright protected material (postmortem tissue images) used with permission of the authors and the University of Iowa's Virtual Hospital, www.vh.org.

specific axons. For example, one may postulate that the superior longitudinal fasciculus (slf), which appears as a long tract connecting the frontal and temporal lobes, is actually an assembly of short-range tracts merging in and exiting out at many points along the tract. DTI cannot distinguish whether this is true or not. Therefore, DTI-based 3D tract reconstruction results should be treated as a macroscopic delineation of tract configurations or a white matter parcellation tool.

CHAPTER 2

Methods

1. Data acquisition

A 1.5 T Philips GyroscanNT was used. DTI data were acquired using a single-shot echo planar imaging (EPI) sequence with parallel imaging (SENSitivity Encoding factor, $R = 2.5$). The imaging matrix was 112×112 with a field of view of 246×246 mm (nominal resolution of 2.2 mm), which was zerofilled to a 256×256 matrix. The image orientation was axial with 2.2 mm slice thickness, which was aligned parallel to the anterior–posterior commissure line. A total of 55 slices covered the entire cerebral hemispheres and the brainstem. The diffusion weighting was encoded along 30 independent orientations with maximum $b = 700$ mm^2/s. Five additional images with minimal diffusion weighting were also acquired. The scanning time of one dataset was about 6 min. In order to enhance the signal-to-noise ratio, the scan was repeated 6 times. A co-registered MPRAGE image (T_1-weighted image) with the same resolution was also recorded for anatomical guidance.

2. Data processing

The DTI datasets were transferred to a PC workstation, re-aligned for bulk motion and processed using DtiStudio (H. Jiang and S. Mori, Johns Hopkins University, http://cmrm.med.jhmi.edu or http://godzilla.kennedykrieger.org). All diffusion-weighted images were visually inspected for apparent artifacts due to subject motion and instrumental malfunction. Then, the 6 elements of the diffusion tensor were calculated for each pixel using multi-variant linear fitting. After diagonalization, three eigenvalues ($\lambda_1, \lambda_2, \lambda_3$) and three eigenvectors (v_1, v_2, v_3) were obtained. For the anisotropy map, the so-called fractional anisotropy (FA) parameter was calculated, which is scaled from 0 (isotropic) to 1 (anisotropic):

$$FA = \sqrt{\frac{1}{2}} \frac{\sqrt{(\lambda_1 - \lambda_2)^2 + (\lambda_2 - \lambda_3)^2 + (\lambda_3 - \lambda_1)^2}}{\sqrt{\lambda_1^2 + \lambda_2^2 + \lambda_3^2}}$$

The eigenvector associated with the largest eigenvalue (v_1) was utilized as an indicator for the fiber orientation. In the color map, red (R), green (G), and blue (B) colors were assigned to right-left, anterior-posterior, and superior-inferior orientations, respectively. For the color presentation, a 24-bit color scheme was used, in which each of RGB colors had an 8-bit (0–255) intensity level. The vector v_1 ($= [v_{1_x}, v_{1_y}, v_{1_z}]$) is a unit vector that always fulfills a condition, $v_{1_x}^2 + v_{1_y}^2 + v_{1_z}^2 = 1$. Intensity values of $v_{1_x} \times 255$, $v_{1_y} \times 255$, to and $v_{1_z} \times 255$ were assigned to the R channel, G channel, and B channel, respectively. In order to suppress orientation information in isotropic brain regions (there should not be a preferential orientation in isotropic areas and the calculated orientations in such areas are dominated by noise), the 24-bit color value was multiplied by the FA value.

3. Three-D tract reconstruction

3.1. Reconstruction algorithm

The 3D tract reconstruction was performed using the FACT (Fiber Assignment by Continuous Tracking) method (Fig. 1), which performs a straightforward linear line propagation based on the v_1 vector angle. An anisotropy threshold of FA > 0.2 was used, while the angle of progression was restricted by using an inner product of more than 0.75 between consecutive eigenvectors.

Fig. 2 shows a schematic diagram of the fiber reconstruction process. Suppose there is a tract with the structure shown in Fig. 2A. DTI measurement provides the fiber orientation information shown in Fig. 2B. For fiber reconstruction, a tract of interest first needs to be identified and marked by a region of interest (ROI) (Bold box). Two possible reconstruction approaches are demonstrated. In the first, the so-called "from-ROI" approach (Fig. 2C), tracking originates from the ROI.

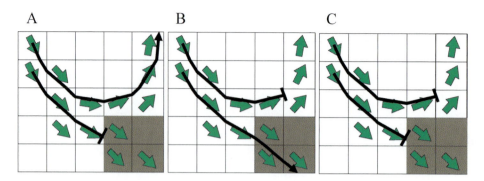

Fig. 1. Schematic diagram of FACT fiber tract reconstruction based on DTI data. Once the fiber orientation (v_1) is estimated at each pixel, putative projections are traced by propagating a line along the estimated fiber orientations. The propagation terminates either when it enters an area with anisotropy lower than a threshold (A: dark boxes) or when the trajectory has a turn judged as too sharp by an inner product between two connected pixels (B). In FACT, both criteria are applied (C).

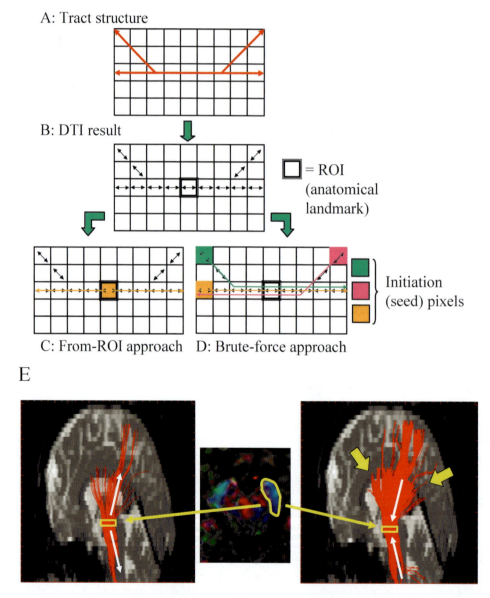

Fig. 2. Principles of tract reconstruction using the "from ROI" and the "brute-force" approaches. (A) Example of a tract structure with 2 branching points. (B) Results of DTI measurement, a vector field that shows the fiber orientation at each pixel. A bold box shows the anatomical landmark where a ROI is defined. (C) Results of the tract reconstruction using the "from ROI" approach, in which tracking is started from the ROI. This generally leads to an incomplete delineation of the tract. (D) Results of the "brute force" approach, in which tracking is started from all pixels. Several initiation (seed) pixels from which the tracking can lead to the same ROI are demonstrated. (E) Results of actual tracking using the cerebral peduncle as a reference ROI. The left and right panels show results for the "from ROI" and "brute force" approaches, respectively.

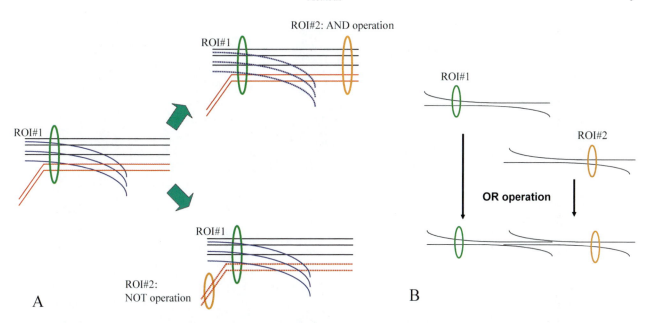

Fig. 3. Schematic diagram of three ROI operations. When the first ROI is drawn, all tracts that penetrate the ROI are retrieved (black, red and blue fibers) (A). If the second ROI is applied as an "AND" operation, only the fibers that penetrate both ROIs are retained (black and red fibers). If a "NOT" operation is applied to the second ROI, a subset of the fibers penetrating the first ROI (ROI #1) but not the second ROI (ROI #2) is selected (black and blue fibers). When the "OR" operation is used, multiple tracking results are combined (B).

In this simple example, the ROI contains only one pixel and, as a result, only 1 line is produced in the "from-ROI" approach, which restricts the labeling to merely a segment of the tract. In the second, so-called "brute-force" approach, all pixels are visited and examined and all tracts originating from such pixels and penetrating the ROI are tracked. This approach is much more time-consuming but leads to a more comprehensive delineation of the tract(s). Fiber reconstruction in this atlas is therefore based on the brute-force approach.

3.2. Editing of reconstruction results using multiple ROIs

In the present atlas, we used several fiber-editing techniques that employ multiple ROIs to delineate tracts of interest. There were three types of multi-ROI operations; AND, OR, and NOT (Fig. 3). These operations were employed diversely depending on the characteristic trajectory of each tract. The most commonly used function was the "AND" operation, applied to identify specific fibers that connect more than one anatomical landmarks depicted by ROIs. The "OR" operation was used for fibers in the limbic system, which tend to have a narrow tubular shape for which a single tract tracking result sometimes failed to delineate their entire length. Use of the "NOT" operation was sometimes necessary to remove a subset of projections from a reconstruction result. As mentioned in the previous section, tracts in this atlas were reconstructed as faithfully as possible based on classical *a priori* anatomical knowledge from postmortem studies. DTI-based fiber tracking often leads to trajectories that have not been well characterized previously. These unexpected tracts might correspond to reality as well as artifact, which presently cannot be validated. In this atlas, such undocumented fibers were removed by the "NOT" operation. The "NOT" operation was also used to remove "relayed" projections often observed in the thalamus. For example, when the anterior thalamic radiation is reconstructed using two ROIs, one at the thalamus and the other at the frontal lobe, a small number of projections that connect the frontal lobe to the thalamus penetrate the thalamus and proceed to various brain regions such as the brainstem and cerebellum. Some of these trajectories are observed reproducibly and could possibly correspond to real connections, but they were chosen to be removed using the "NOT" operation as they apparently do not belong to the classical description of the anterior thalamic radiation.

3.3. ROI drawing strategy

Once all the data have been preprocessed, the tracts have to be reconstructed by proper choice of reference ROIs and subsequently assigned. In Fig. 4, an example of actual fiber reconstruction of the inferior fronto-occipital fasciculus using multiple ROIs is shown. These ROIs are strategically located based on known fiber trajectories. From this example, it can be seen that multiple ROIs pose strong constraint and large ROIs that surround the target areas give rise to clean and reproducible definition of tracts of interest. Actual locations and sizes of ROIs used in this book followed our previous publication for fibers in the brainstem (Stieltjes et al., 1999) as well as in the cerebral hemispheres (Mori, 2002).

3.4. Limitations of DTI-based reconstruction

Compared to invasive chemical-tracer based techniques, DTI-based tract reconstruction has several important limitations. First of all, it does not distinguish afferent from efferent projections. Secondly, because of a limitation in the spatial

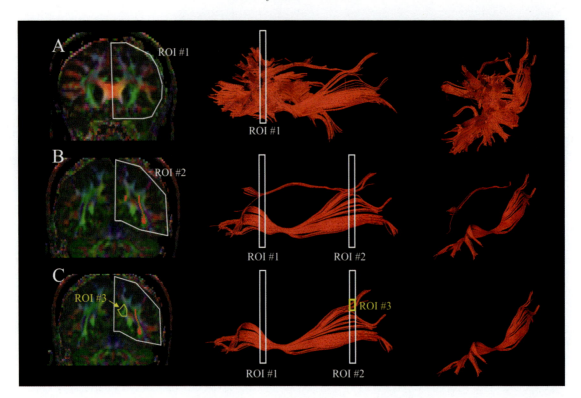

Fig. 4. An example of fiber editing using the "AND" and "NOT" operations to reconstruct the inferior fronto-occipital fasciculus (IFO). First, a large ROI that delineates the entire frontal lobe is drawn at a coronal slice level. This leads to a large number of trajectories that penetrate the ROI (A), including callosal connections to the contralateral hemisphere, the anterior thalamic radiation, and various association fibers. The second ROI is drawn at the occipital lobe (B) and an "AND" operation is used to search only for trajectories that penetrate both ROIs. This leads to delineation of the IFO. However, it turns out that a part of the cingulum also penetrates the same ROIs. This cingulum component is removed by a "NOT" operation (C, yellow ROI), which results in an IFO representation that is similar to that of classical drawings by neuroanatomists.

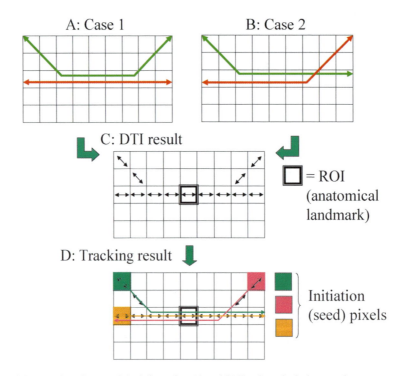

Fig. 5. Schematic diagram of degeneration of connectivity information. (A and B) Two hypothetical cases of tract structures consisting of two axonal tracts. (C) The vector map obtained from DTI measurement. (D) Results of "brute force" tract reconstruction.

resolution, which is typically 2–3 mm for each pixel dimension, specific connectivity information of individual axons degenerates. This point is illustrated in Fig. 5, where two hypothetical structures of tract systems leading to similar DTI results are shown. The "brute force" approach can reveal the overall configuration of the tract by collectively displaying all tracking results, but each tracking result (e.g., green, pink, or orange line in Fig. 5D) does not necessarily represent a real axonal connectivity.

The third important limitation occurs when there are more than 2 tract families with different orientations within a pixel. For example, if there is a dominant population of one tract mixed with a small number of tracts bearing different fiber orientations, the DTI results can provide only the pathway for the dominant fibers. If there are two equal populations of tracts with different orientations within a pixel, the pixel tends to have low anisotropy, as previously discussed (Fig. 5, Chapter 1). This leads to a termination of the tracking before delineating the entire length of the tract. For example, low anisotropy areas due to merging of projection fibers and the corpus callosum are shown in Fig. 5, Chapter 1. Because of these areas of low anisotropy, tracking results of the projection tracts in Fig. 2E terminate prematurely before they reach the cortical areas (indicated by yellow arrows). This type of possible "lack of labeling" is an important limitation one needs to know.

4. Three-dimensional volume definition of gray matter structures

Using the co-registered MPRAGE images, the ventricles, caudate, lentiform nucleus, thalamus, and hippocampus (including the amygdala) were manually defined to visualize their spatial relationship with the various white matter tracts three-dimensionally.

5. Visualization

5.1. Three-D visualization

The trajectories of the white matter tracts and the volumes of gray matter nuclei were visualized using Amira (Mercury Computer System Inc., San Diego).

5.2. 2D presentation

In this atlas, two types of 2D presentations were used: DTI-based color-coded orientation maps (color maps) and white matter parcellation maps. In the latter, 3D reconstruction results are superimposed on T_1-based anatomical maps (MPRAGE) (Fig. 6).

The color map is created directly from the diffusion tensor vector data and various tracts can be identified based on the orientation information. Because more and more clinical MRI scanners with DTI capability are becoming available and because the color map contains a great amount of white matter anatomical information, we anticipate it will be an important clinical tool. However, interpretation of color maps is far from trivial. Two different tract systems may have the same color if their in-plane fiber orientations are the same, while the same tract system may change color because it changes orientation. In the white matter parcellation map, white matter tracts are color-coded specifically for each tract. As a result, different tracts have different colors, simplifying the interpretation. However, clinical application of this approach is less straightforward, because the parcellation map requires 3D tract reconstruction, the result of which depends on ROI placement. The tracking result is also affected by image resolution and signal-to-noise ratio. For reproducible generation of parcellation maps, rather strict protocols for imaging and tract reconstruction must be defined and followed.

To generate this atlas, 3D coordinate information of each tract was superimposed on MPRAGE images to create the parcellation maps. Even though the SENSE acquisition scheme drastically reduced B_0-related image distortions, a small amount of mismatch between the DTI and MPRAGE remained around the nasal-sinus areas and, thus, care must be taken to interpret data in these regions.

5.3. Nomenclature and annotation

Nomenclature for the white matter is sometimes confusing. Probably most straightforward is the assignment of fibers that include the two connected anatomical regions in their names (e.g., corticospinal tract, corticopontine tract, inferior fronto-occipital fasciculus), but not all tracts follow this nomenclature (e.g., fornix, superior longitudinal fasciculus, uncinate fasciculus). There is also a group of names that is not necessarily related to specific tract connectivity. One example of this is the internal capsule (Fig. 7), which has subdivisions (anterior limb, genu, posterior limb, and retrolenticular part) based on anatomical locations. These areas collectively contain corticoafferent corticopetal pathways, such as thalamo-cortical tracts, as well as corticoefferent (corticofugal) pathways such as corticothalamic, corticoreticular, corticopontine, corticobulbar, and corticospinal tracts.

In this atlas, many white matter tracts were reconstructed as faithfully as possible using multiple regions of interest chosen based on existing anatomical knowledge. Based on the 3D trajectory information, various structures were identified in the color maps and annotated at the right side (see Fig. 6). On the left side, names of white matter structures that were not reconstructed or not related to single specific tracts were annotated using white color. These include (see Fig. 6, Chapter 1, for visual support):

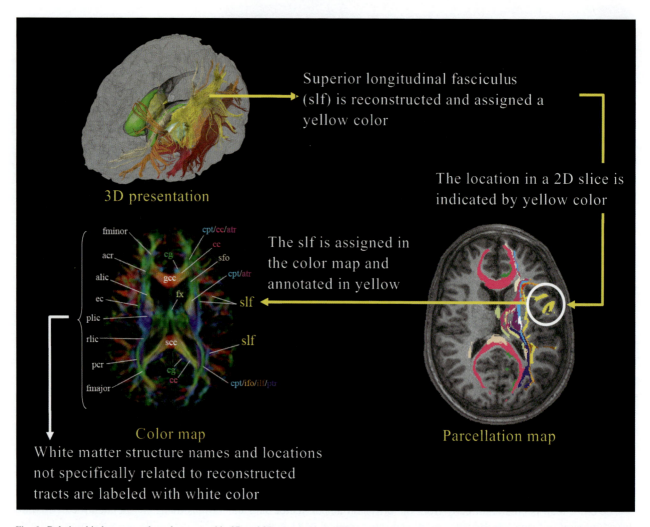

Fig. 6. Relationship between color schemes used in 3D and 2D presentations. White matter tracts are reconstructed three-dimensionally and visualized using colors assigned to each tract. In the 2D parcellation map, the locations of these reconstructed tracts at a given slice level are indicated by the same color. By comparing the parcellation map and color map, visible structures in the color map are assigned and annotated. For the annotation, the same color as used for the 3D presentation (and parcellation map) is applied for each tract visible in a particular slice.

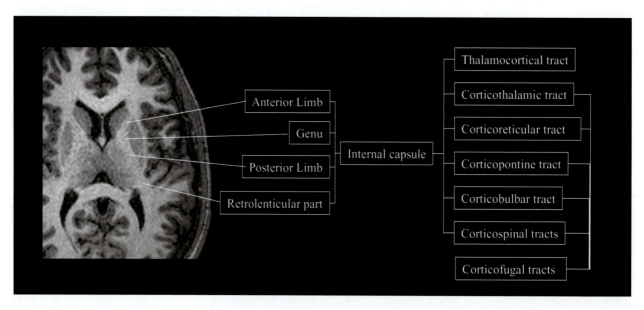

Fig. 7. Relationship between various nomenclatures of white matter tracts in the internal capsule. Names in the left column are related to anatomical locations while those in the right column are connectivity-based.

Corona radiata:
This structure contains the corticopetal (corticoafferent: thalamocortical fibers), long corticofugal (corticoefferent: corticothalamic, corticostriatal, corticopontine, corticobulbar, corticospinal and corticoreticular), and commissural fibers. In the atlas, it is divided into anterior (acr), superior (scr), and posterior (pcr) parts, although the boundaries are arbitrarily defined.

Internal capsule:
The corona radiata converges into the internal capsule, which contains the thalamocortical and long corticofugal tracts. It can be divided into the anterior limb (alic), posterior limb (plic) and retrolenticular part (rlic), as shown in Fig. 7.

Cerebral peduncle:
At the midbrain level, the internal capsule converges into the cerebral peduncle, which contains the long corticofugal tracts.

Sagittal stratum:
This structure seems contiguous to the posterior region of the corona radiata (Fig. 6, Chapter 1), its main constituent being the optic radiation (a part of the thalamocortical tract). It also contains commissural fibers from the splenium of the corpus callosum and corticofugal fibers as well as association fibers, most notably the inferior fronto-occipital fasciculus and the inferior longitudinal fasciculus.

External capsule:
Seen as a thin layer of white matter lateral to the lentiform nucleus, it contains association fibers such as the superior longitudinal fasciculus, inferior fronto-occipital fasciculus, and uncinate fasciculus. With current imaging resolution for DTI, the extreme and external capsules can not be discriminated.

CHAPTER 3

Three-Dimensional Atlas of Brain White Matter Tracts

White matter tracts were classified into four groups: tracts in the brainstem and projection, association, and commissural tracts in the cerebral hemispheres. Projection fibers connect cortical and subcortical gray matter, association tracts connect two cortical areas, and commissural tracts connect right and left hemispheres. In the following sections, 3D architectures of these tracts are reconstructed.

1. Tracts in the brainstem

Five major white matter tracts were reconstructed in the brainstem. These are: the superior, middle, and inferior cerebellar peduncles, the corticospinal tract and the medial lemniscus. A schematic diagram and 3D trajectories of these tracts are shown in Fig. 1. The global trajectories of these tracts have been described in many neuroanatomy textbooks and the DTI-based fiber tracking could faithfully reconstruct them. Unlike histology-based techniques, the MRI-based reconstruction allows viewing of the structures from multiple orientations and adding or removing structures of interest as desired (Fig. 2 and 3).

The superior cerebellar peduncle (scp) is the main efferent pathway from the dentate nucleus of the cerebellum toward the thalamus (Fig. 4). The tract reconstruction result shows that the one side of the superior cerebellar peduncle originates at the dentate nuclei and the other at the thalamus. In 2D slices, it can be discretely identified in an axial slice at the level of the dentate nucleus (slice #2 in Fig. 4). At a superior level (slice #1 and #4 in Fig. 4), its decussation can be found as a circular red (right–left orientation) structure. Although the majority of the fibers crosses the midline at the decussation, the reconstructed trajectory remains in the ipsilateral side. This is due to limitation of the fiber tracking technique, which tends to provide > < (kissing) solution when two fibers have X (crossing) trajectories.

The inferior cerebellar peduncle (icp) contains afferent and efferent connections to the cerebellum (Fig. 5). It originates in the caudal medulla, traverses the pons, and branches into the cerebellar cortex. In the color map, it can be readily identified at the dorsal area of the medulla (slice #5 in Fig. 5) and the pons (slice #3 and #4 in Fig. 5).

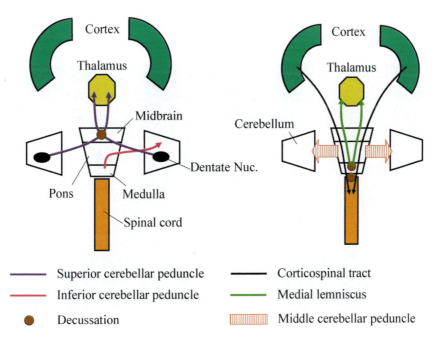

Fig. 1. Schematic diagram of the tracts in the brainstem.

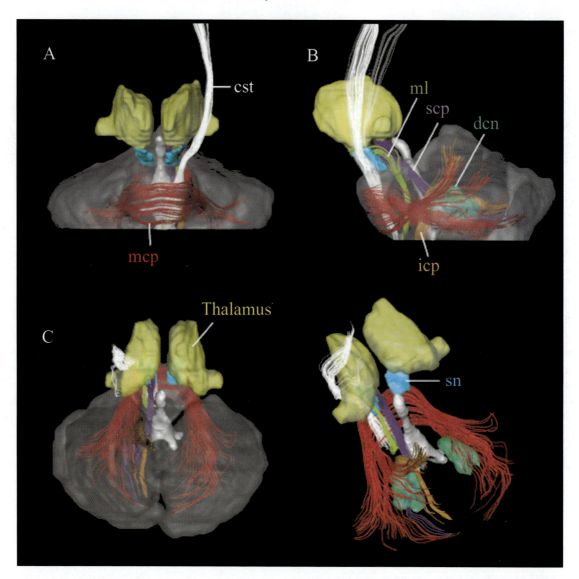

Fig. 2. 3D reconstruction results of 5 major tracts in the brainstem. Tracts are viewed from the anterior (A), left (B), superior (C), and oblique (left-superior-posterior) (D) orientations. The cerebellum is represented by a semi-transparent gray structure in A, B and C. The cerebral aqueduct and the IV ventricle are represented by a white 3D reconstructed structure. Abbreviations are: cst: corticospinal tract; dcn: deep cerebellar nuclei (dentate nucleus); icp: inferior cerebellar peduncle; mcp: medial cerebellar peduncle; ml: medial lemniscus; scp: superior cerebellar peduncle; sn: substantia nigra.

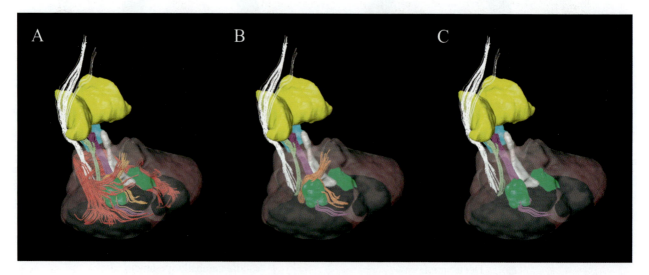

Fig. 3. Examples of structure stripping to reveal inner structures. In B, the mcp was excluded, allowing better visualization of the cst, ml, icp and scp. In C, the icp was removed, allowing better assessment of the relationship between the scp, dnc and IV ventricle.

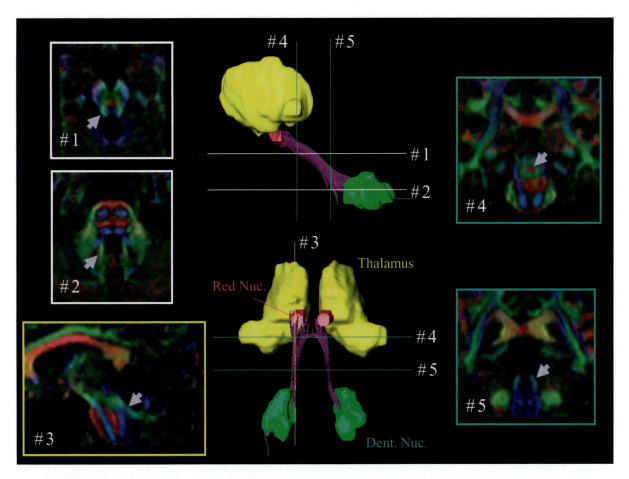

Fig. 4. The trajectory of the superior cerebellar peduncle and its identification in color maps at various slice locations and orientations. Locations of 2D slices are indicated in the 3D reconstruction viewed from the left (upper panel) and inferior (bottom panel) orientations.

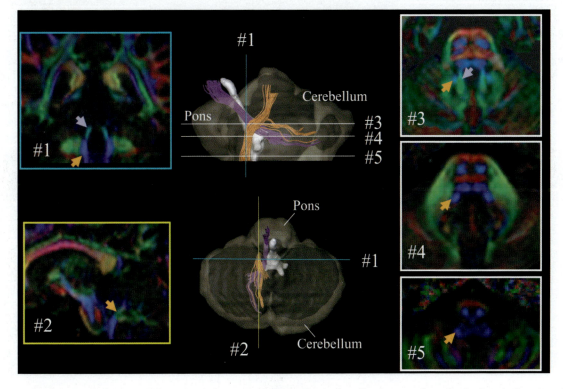

Fig. 5. The trajectory of the inferior cerebellar peduncle (orange) and its identification in color maps at various slice levels and orientations. The icp is indicated by orange arrows and the scp by purple arrows. The semi-transparent gray object is the cerebellum and the white structure includes the cerebral aqueduct and the IV ventricle. Locations of 2D slices are indicated in the 3D reconstruction viewed from the left (upper panel) and superior (bottom panel) orientations.

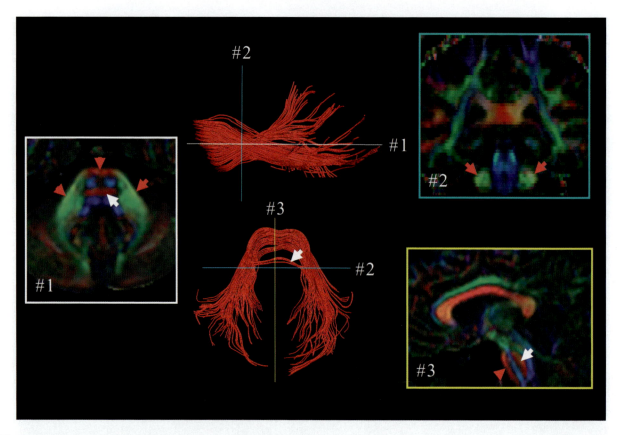

Fig. 6. The trajectory of the middle cerebellar peduncle and its identification in color maps (red arrows) at three slice locations and orientations. Locations of the 2D slices are indicated in the 3D reconstruction view from the left (upper panel) and superior (bottom panel) orientations. The transverse pontine fibers (white arrows) are part of the mcp.

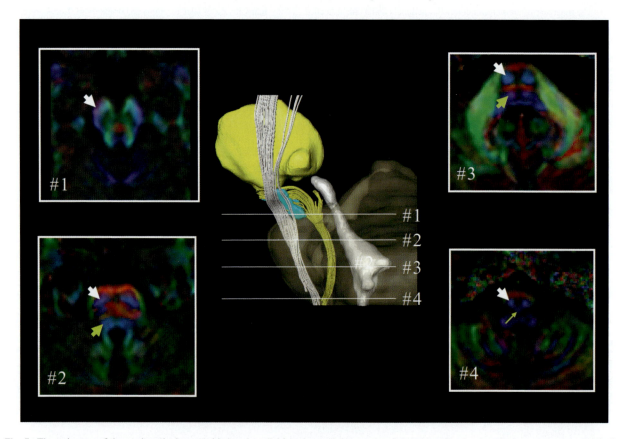

Fig. 7. The trajectory of the corticospinal tract (white) and medial lemniscus (light green) and their identification in color maps at various axial slice levels. White and light green arrows in the color maps indicate the locations of the cst and ml, respectively. The levels of the 2D slices are indicated in the 3D reconstruction viewed from the left side. The thalamus (yellow), substantia nigra (blue), ventricle (white), and cerebellum (gray) are also shown in the 3D reconstruction.

The middle cerebellar peduncle (mcp) also contains efferent fibers from the pons to the cerebellum (pontocerebellar tracts), forming a massive sheet-like structure that wraps around the pons (Fig. 6). The cross section of this tract can be readily identified in coronal planes (slice #2 in Fig. 6) through the pons. The transverse pontine fibers indicated by white arrows in slices #1 and #3 in Fig. 6 are also part of pontocerebellar tracts.

The medial lemniscus (ml) is a major pathway for ascending sensory fibers (Fig. 7). It decussates at the level of the ventral medulla (slice #4 in Fig. 7). As it ascends the pons, it travels along the dorsal aspect of the pons and midbrain. The right and left tracts can be identified at slice levels #2 and #3 and its decussation at the slice #4 in Fig. 7. At the midbrain level (slice #1), it becomes too dispersed for discrete identification.

The corticospinal tract (cst) is a descending pathway from the cortex; it penetrates the cerebral peduncle (indicated by the white arrow in slice #1) and is readily identifiable in the pons and medulla (slices #2–#4).

2. Projection fibers

Two classes of projection fibers were reconstructed for this atlas: the corticothalamic/thalamocortical fibers (collectively called thalamic radiations) and the long corticofugal (corticoefferent) fibers. The corticofugal fibers include such fibers as the corticopontine, corticoreticular, corticobulbar, and corticospinal tracts. A schematic diagram of their trajectories is shown in Fig. 8.

In Fig. 9, the 3D trajectories of these reconstructed projection fibers are shown. They all penetrate the internal capsule either between the thalamus and the putamen or between the caudate and the putamen. When approaching to the cortex, they fan out to form the corona radiata (see also Fig. 6 in Chapter 1).

In Fig. 10, the locations of these projection fibers are indicated in several 2D slices. The thalamus is known to have reciprocal connections to a wide area of the cortex (see Fig. 10C). In this atlas, the different parts of the thalamic radiation that penetrate the anterior limb, posterior limb, and retrolenticular part of the internal capsule are labeled as anterior, superior (central), and posterior thalamic radiations, respectively. The posterior thalamic radiation should include the optic radiation, which connects the lateral geniculate nucleus and the occipital lobe. Although these three components of the thalamic radiation are differently color-coded in this reconstruction, there is no precise distinction between the anterior and superior or superior and posterior thalamic radiations and, thus, their boundaries should not be considered as definite for assignment of these fibers.

The internal capsule also contains descending pathways from the cortex (long corticofugal pathways) toward the brainstem. The vast majority of these pathways terminate in some pontine nuclei and are called corticopontine tracts (light blue in Fig. 10). Depending upon their area of origin in the cortex, the corticopontine tracts are subdivided into fronto, parieto, occipito, and temporo-pontine tracts. The pontine nuclei then send projections to the contralateral cerebellum through the

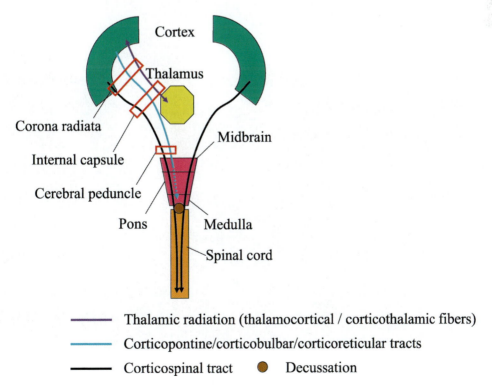

Fig. 8. A schematic diagram of trajectories of projection fibers reconstructed in this atlas. The decussation is that of the corticospinal tract.

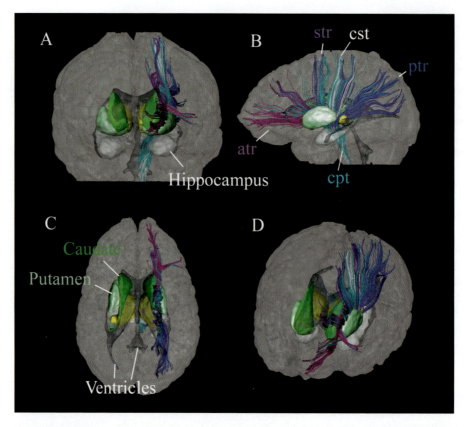

Fig. 9. 3D reconstruction results of projection fibers. Tracts are viewed from the anterior (A), left (B), superior (C), and oblique (left-superior-anterior) (D) orientations. The hemispheres are delineated in semi-transparent gray. Abbreviations are: atr: anterior thalamic radiation; cpt: corticopontine tract; cst: corticospinal tract; ptr: posterior thalamic radiation; str: superior thalamic radiation.

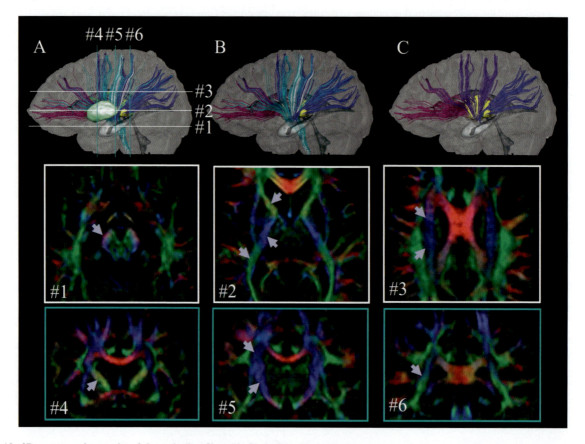

Fig. 10. 3D reconstruction results of the projection fibers (A–C). In (B), the putamen was removed to better visualize the tracts, and in (C), the corticopontine (light blue) and corticospinal (white) tracts were removed for better assessment of the thalamic radiations. Locations of the projection tracts are indicated by purple arrows in the 6 color maps.

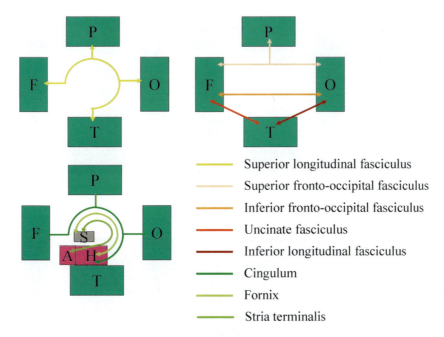

Superior longitudinal fasciculus

Superior fronto-occipital fasciculus

Inferior fronto-occipital fasciculus

Uncinate fasciculus

Inferior longitudinal fasciculus

Cingulum

Fornix

Stria terminalis

Fig. 11. A schematic diagram of some cortico-cortical connections by association fibers. Abbreviations are F: frontal; P: parietal; O: occipital, and T: temporal cortices; A: amygdala; H: hippocampus; S: septal area.

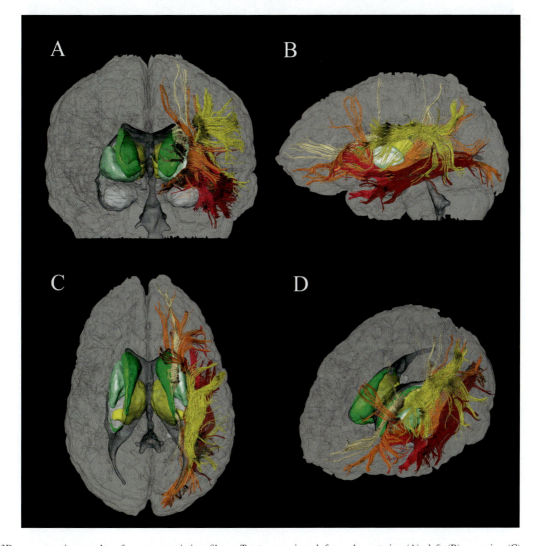

Fig. 12. 3D reconstruction results of some association fibers. Tracts are viewed from the anterior (A), left (B), superior (C), and oblique (left-antero-superior) (D) orientations. Color coding: superior longitudinal fasciculus is yellow, inferior fronto-occipital fasciculus is orange, uncinate fasciculus is red and inferior longitudinal fasciculus is brown. Cerebral hemispheres are delineated by semi-transparent gray. Thalami are yellow, ventricles are gray, caudate nuclei are green and lentiform nuclei are light green.

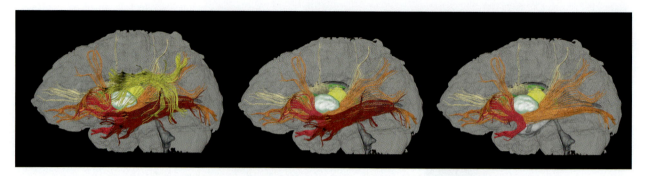

Fig. 13. Structural stripping to reveal inner structures. In the middle and right panels, the superior longitudinal fasciculus (yellow) and inferior longitudinal fasciculus (brown) are removed sequentially, revealing the uncinate (red) and inferior fronto-occipital (orange) fasciculus.

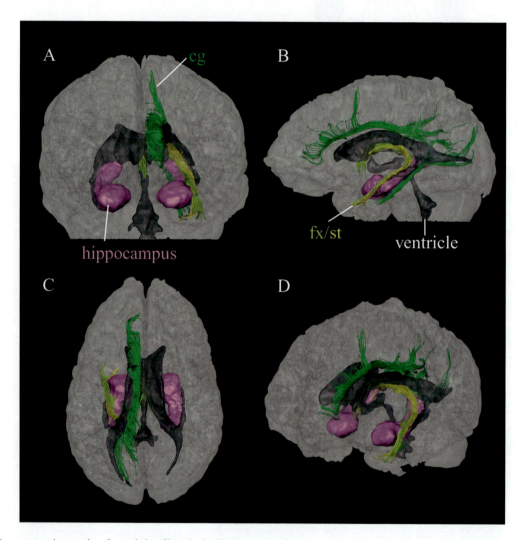

Fig. 14. 3D reconstruction results of association fibers in the limbic system. Tracts are viewed from the anterior (A), left (B), superior (C), and oblique (left-anterior) (D) orientations. Abbreviations are: cg: cingulum; fx: fornix, and st: stria terminalis. The cerebral hemispheres are delineated in semi-transparent gray. The ventricular system is depicted in gray and the hippocampi and amygdalae, in purple.

middle cerebral peduncle, forming the massive cortico-ponto-cerebellar pathway. The corticopontine tracts pass through both the internal capsule (slice #2 in Fig. 10) and the cerebral peduncle (slice #1 in Fig. 10). Strictly speaking, this reconstruction also contains other cortico-brainstem pathways such as the corticoreticular and corticobulbar tracts. Because DTI-based tract reconstruction cannot differentiate these pathways from the corticopontine tracts, they were included within the corticopontine tracts.

The middle portion of the cerebral peduncle is believed to contain tracts that pass through the brainstem and reach the spinal cord, it is therefore called the corticospinal tract (white in Fig. 10). In this tract, descending projections from the precentral (motor) area are the dominant constituent, although some branches from the somatosensory and parietal cortices are also included.

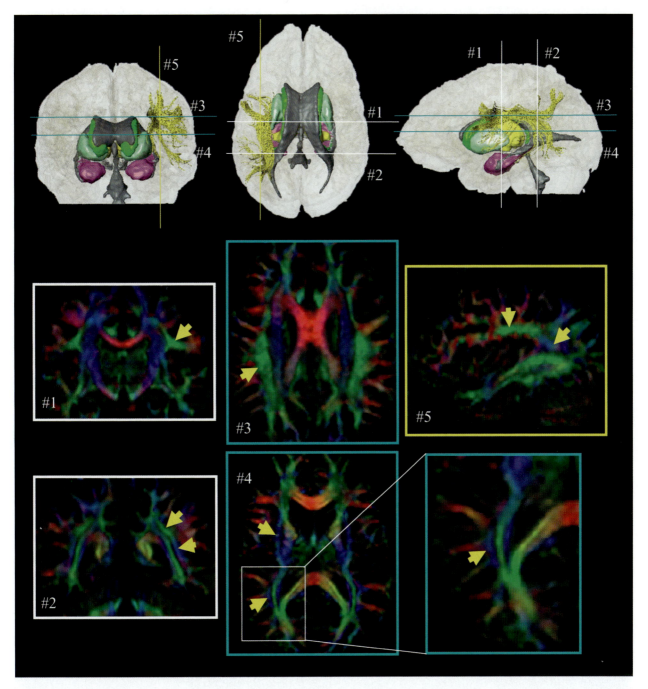

Fig. 15. The trajectory of the superior longitudinal fasciculus and its identification in the color maps at various slice levels and orientations. Locations of 2D slices are indicated in the 3D panels. The hippocampus and amygdala (purple), thalamus (yellow), ventricular system (gray), caudate (green), and putamen (light green) are also shown in the 3D reconstruction.

3. Association fibers

Association fibers connect different areas of the cortex and are classified into short and long association fibers. The former connect areas within the same lobe and include the fibers connecting adjacent gyri, called the U-fibers. The long association fibers connect different lobes, forming prominent fiber bundles. Fig. 11 shows a schematic diagram of the trajectories of some association tracts. Their approximate locations and trajectories have been well documented in the literature. In the present atlas, these major long association fibers were reconstructed based on previous anatomical descriptions. These include the superior longitudinal fasciculus (slf), inferior longitudinal fasciculus (ilf), superior fronto-occipital fasciculus (sfo), inferior fronto-occipital fasciculus (ifo), and uncinate fasciculus (unc) (Fig. 12 and Fig. 13). Three major association fibers connecting the limbic system, namely the cingulum (cg), fornix (fx), and stria terminalis (st), are also presented in Fig. 14.

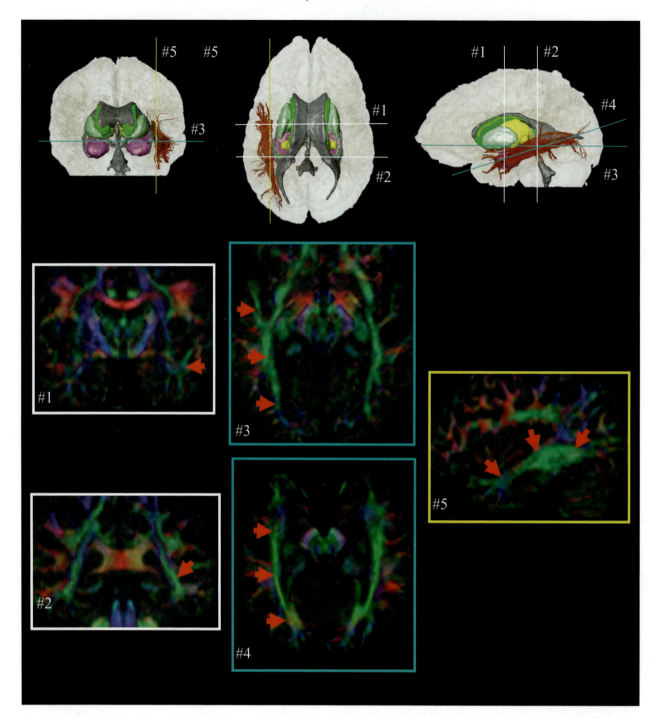

Fig. 16. The trajectory of the inferior longitudinal fasciculus and its identification in color maps at various slice levels and orientations. Locations of the 2D slices are indicated in the 3D panels. The hippocampus and amygdala (purple), thalamus (yellow), ventricular system (gray), caudate (green), and putamen (light green) are also shown in the 3D reconstruction.

Superior longitudinal fasciculus

This tract is located at the supero-lateral side of the putamen and forms a large arc (also called the arcuate fasciculus), sending branches into the frontal, parietal, occipital, and temporal lobes. The tract can be easily located in many coronal slices as a large and intense (high anisotropy) fiber bundle running along the superior edge of the insula (Fig. 15).

Inferior longitudinal fasciculus, inferior fronto-occipital fasciculus, and uncinate fasciculus

If a coronal slice is chosen at the retrolenticular part of the internal capsule, the sagittal stratum (green fiber bundle indicated by a red arrow in Fig. 16, slice #2) can be identified in the color map, which contains two large association fibers. These are the inferior longitudinal fasciculus (Fig. 16), connecting the occipital and temporal lobes, and the inferior fronto-occipital fasciculus (Fig. 17), connecting the frontal and occipital lobes. The inferior fronto-occipital fasciculus has

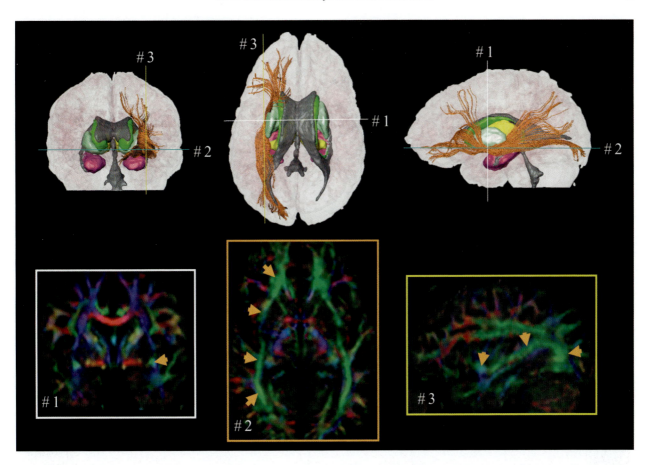

Fig. 17. The trajectory of the inferior fronto-occipital fasciculus and its identification in color maps at various slice levels and orientations. Locations of the 2D slices are indicated in the 3D panels. The hippocampus and amygdala (purple), thalamus (yellow), ventricular system (gray), caudate (green), and putamen (light green) are also shown in the 3D reconstruction.

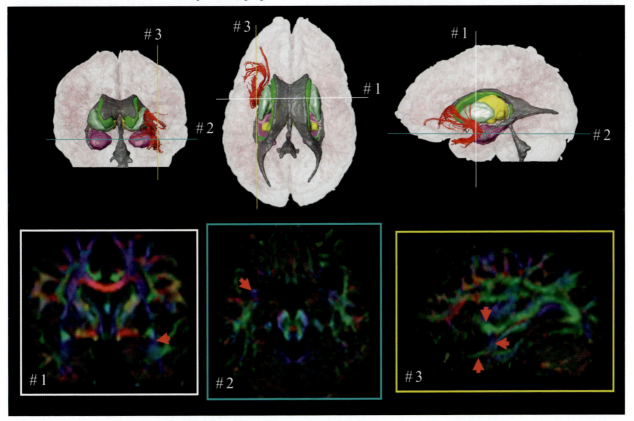

Fig. 18. The trajectory of the uncinate fasciculus and its identification in color maps at various slice levels and orientations. Locations of the 2D slices are indicated in the 3D panels. The hippocampus and amygdala (purple), thalamus (yellow), ventricular system (gray), caudate (green), and putamen (light green) are also shown in the 3D reconstruction.

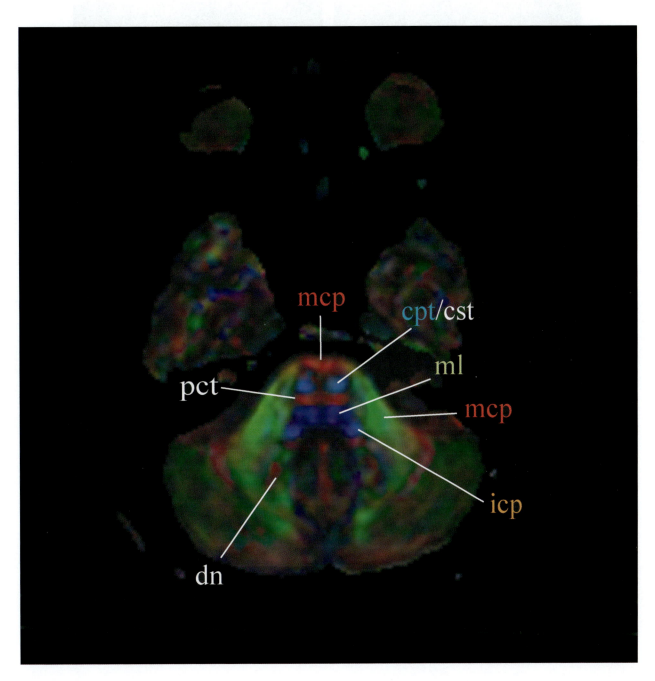

cpt: corticopontine tract
cst: corticospinal tract
dn: dentate nucleus
icp: inferior cerebellar peduncle

mcp: middle cerebellar peduncle
ml: medial lemniscus
pct: pontine crossing tract

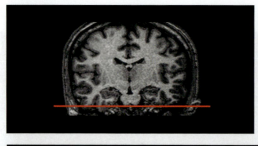
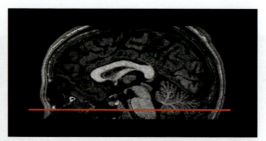
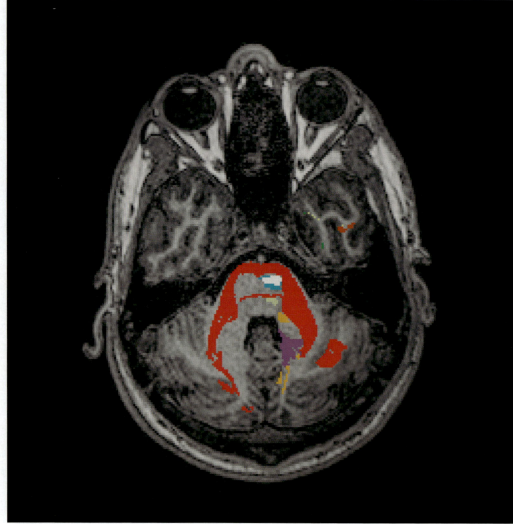

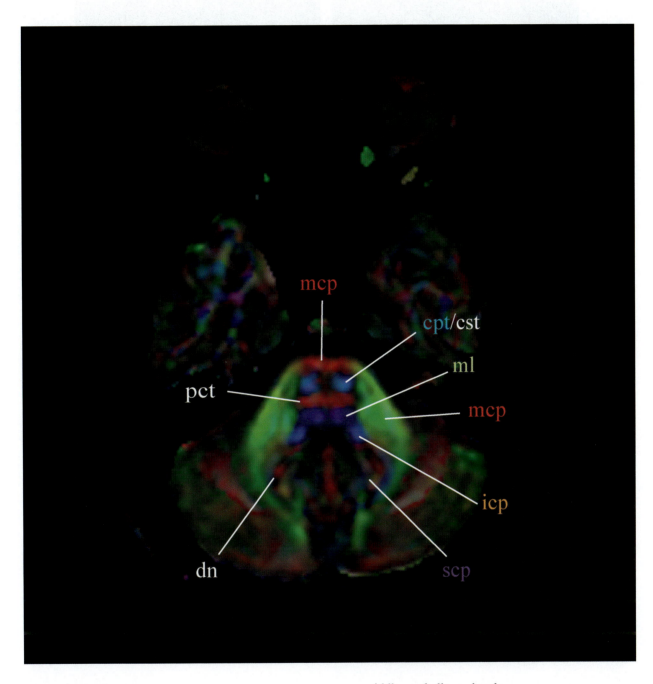

cpt: corticopontine tract
cst: corticospinal tract
dn: dentate nucleus
icp: inferior cerebellar peduncle

mcp: middle cerebellar peduncle
ml: medial lemniscus
pct: pontine crossing tract
scp: superior cerebellar peduncle

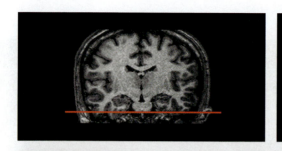
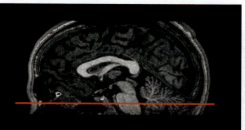
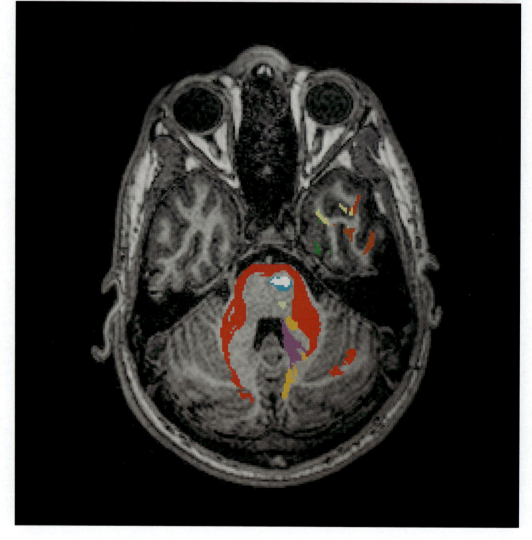

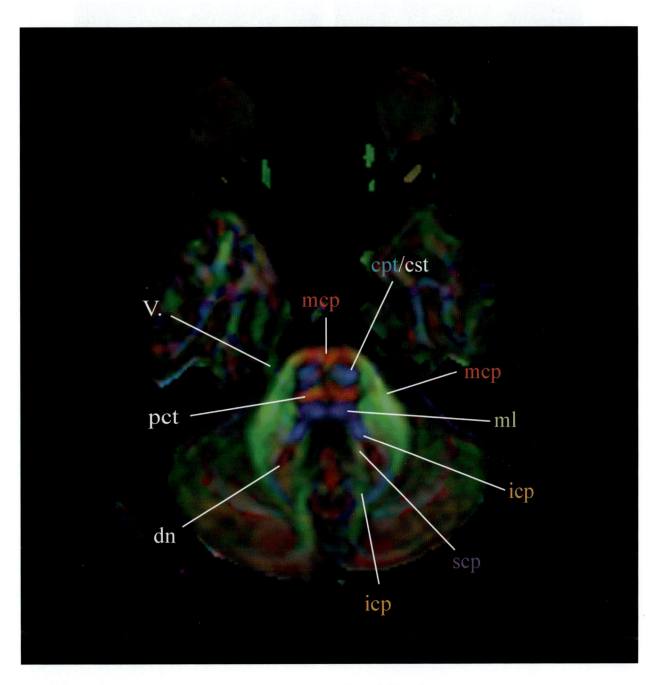

cpt: corticopontine tract
cst: corticospinal tract
dn: dentate nucleus
icp: inferior cerebellar peduncle
mcp: middle cerebellar peduncle

ml: medial lemniscus
pct: pontine crossing tract
scp: superior cerebellar peduncle
V.: Fifth cranial nerve (trigeminal nerve)

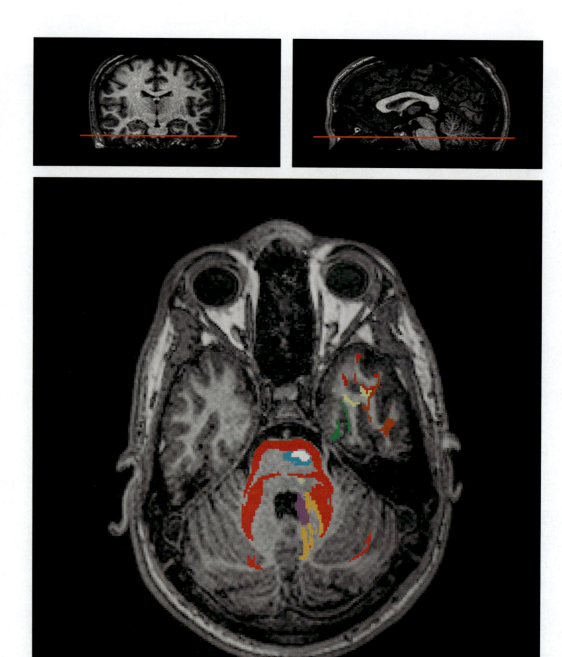

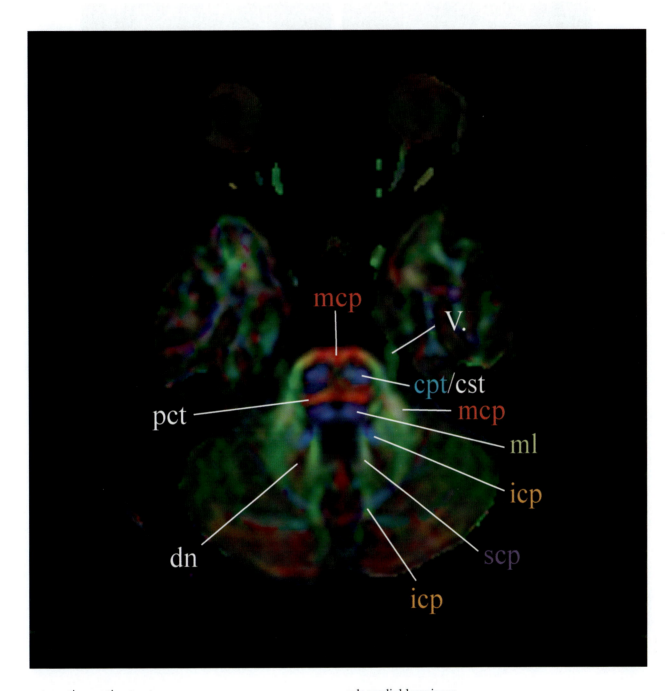

cpt: corticopontine tract
cst: corticospinal tract
dn: dentate nucleus
icp: inferior cerebellar peduncle
mcp: middle cerebellar peduncle

ml: medial lemniscus
pct: pontine crossing tract
scp: superior cerebellar peduncle
V.: Fifth cranial nerve (trigeminal nerve)

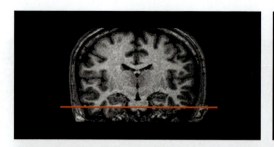
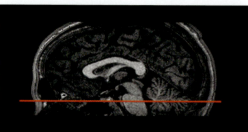

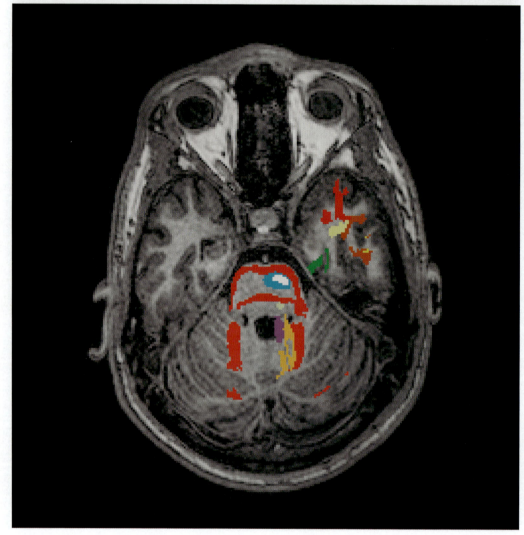

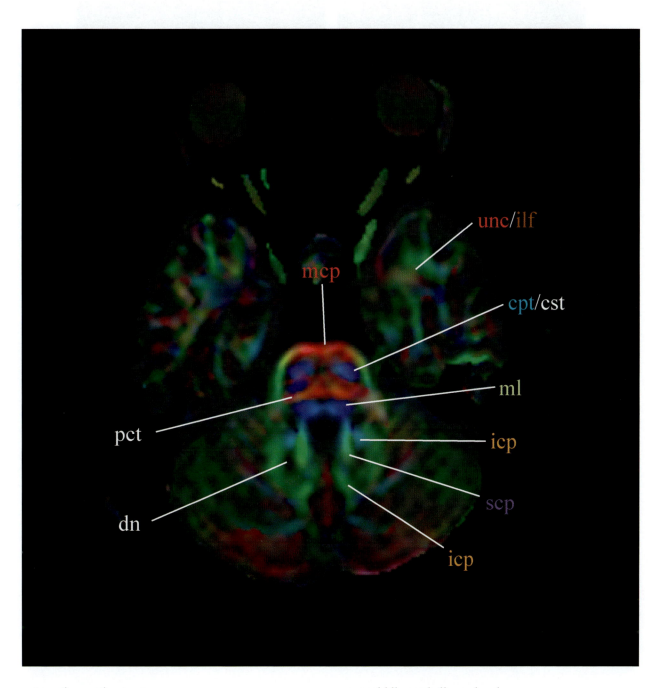

cpt: corticopontine tract
cst: corticospinal tract
dn: dentate nucleus
icp: inferior cerebellar peduncle
ilf: inferior longitudinal fasciculus

mcp: middle cerebellar peduncle
ml: medial lemniscus
pct: pontine crossing tract
scp: superior cerebellar peduncle
unc: uncinate fasciculus

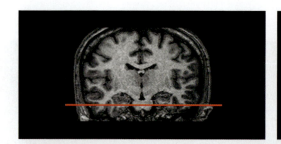
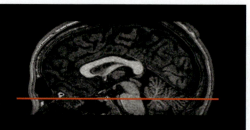

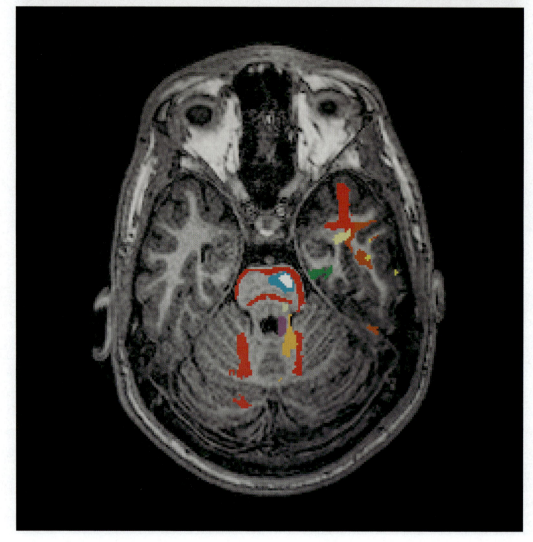

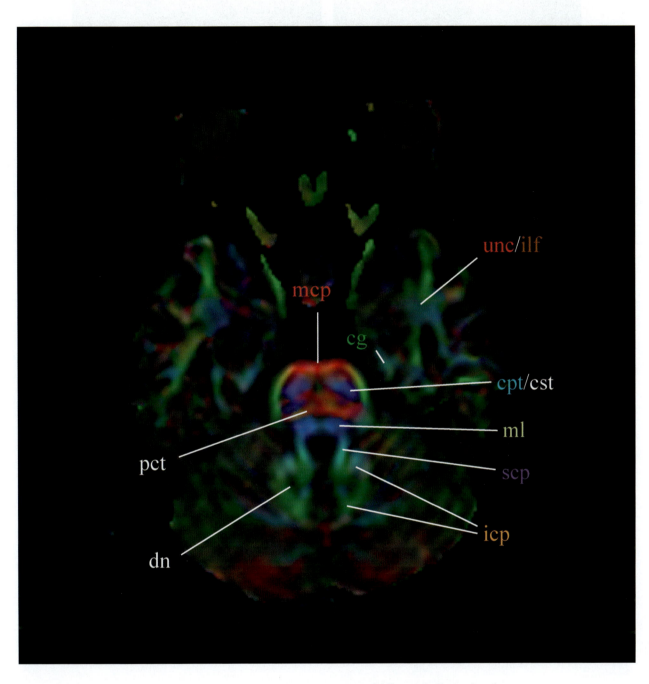

cg: cingulum
cpt: corticopontine tract
cst: corticospinal tract
dn: dentate nucleus
icp: inferior cerebellar peduncle
ilf: inferior longitudinal fasciculus

mcp: middle cerebellar peduncle
ml: medial lemniscus
pct: pontine crossing tract
scp: superior cerebellar peduncle
unc: uncinate fasciculus

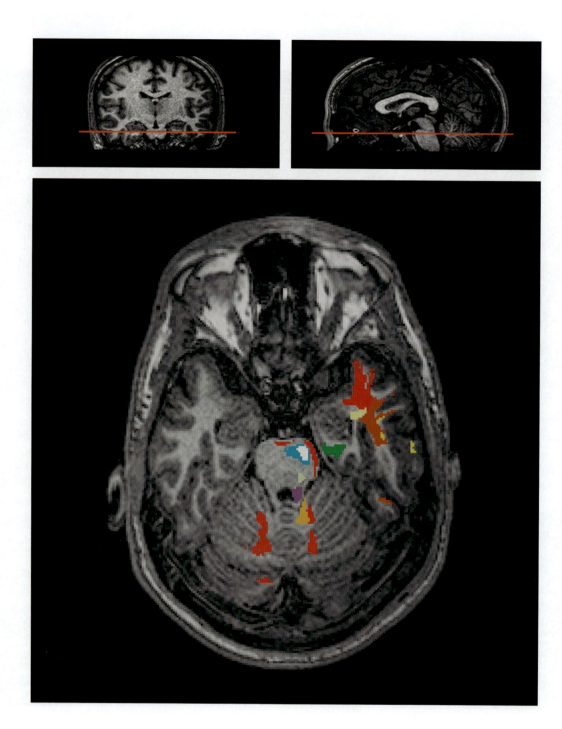

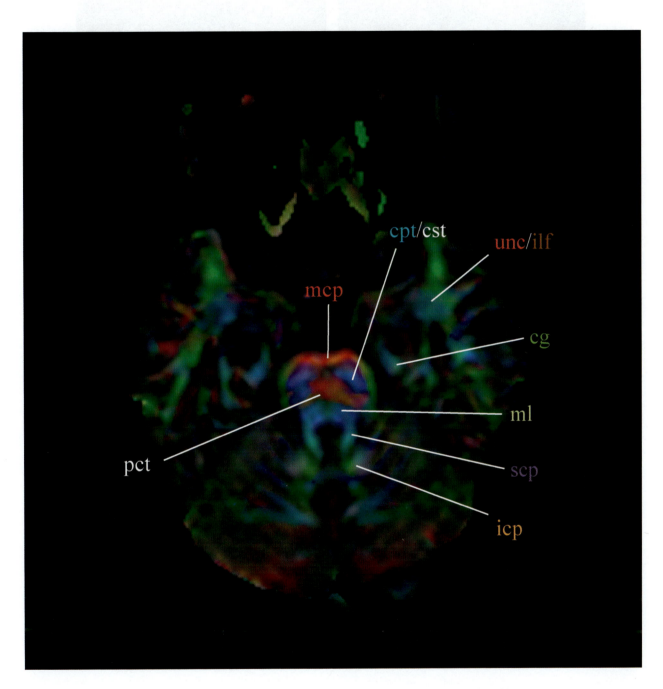

cg: cingulum
cpt: corticopontine tract
cst: corticospinal tract
icp: inferior cerebellar peduncle
ilf: inferior longitudinal fasciculus

mcp: middle cerebellar peduncle
ml: medial lemniscus
pct: pontine crossing tract
scp: superior cerebellar peduncle
unc: uncinate fasciculus

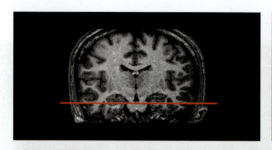
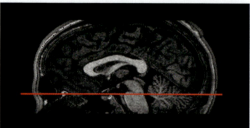
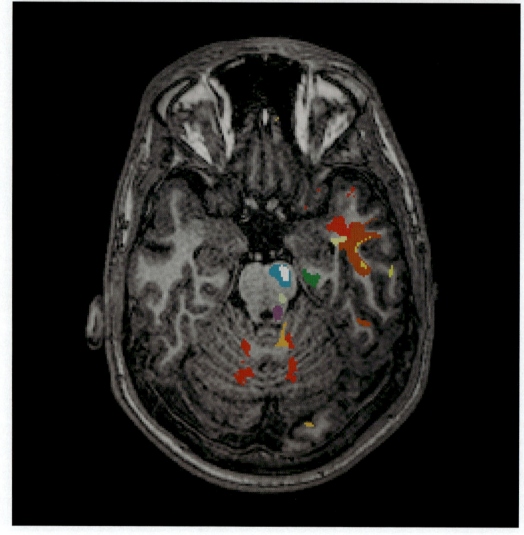

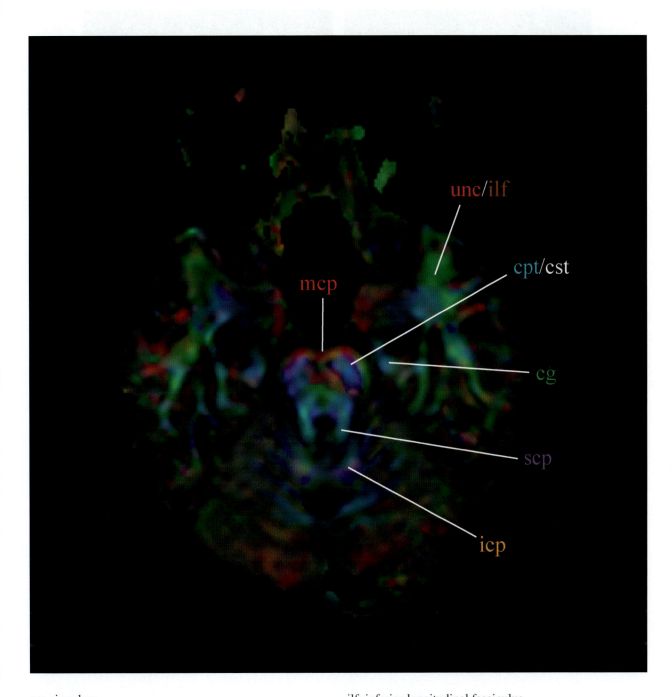

cg: cingulum
cpt: corticopontine tract
cst: corticospinal tract
icp: inferior cerebellar peduncle

ilf: inferior longitudinal fasciculus
mcp: middle cerebellar peduncle
scp: superior cerebellar peduncle
unc: uncinate fasciculus

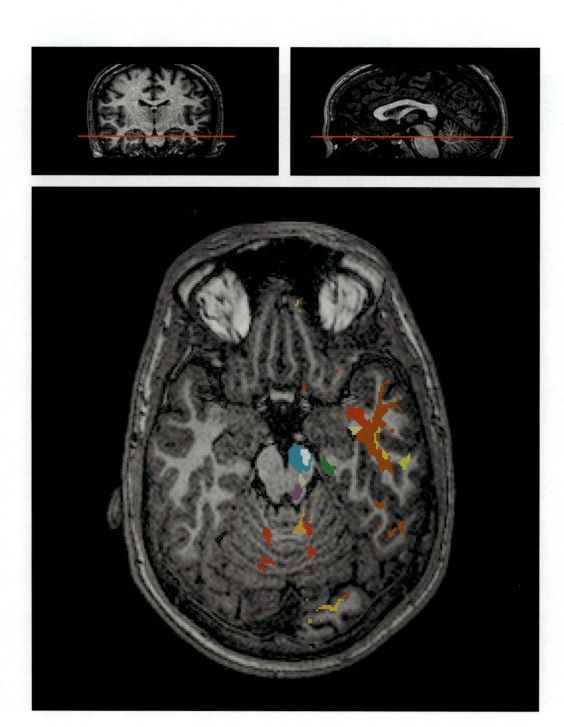

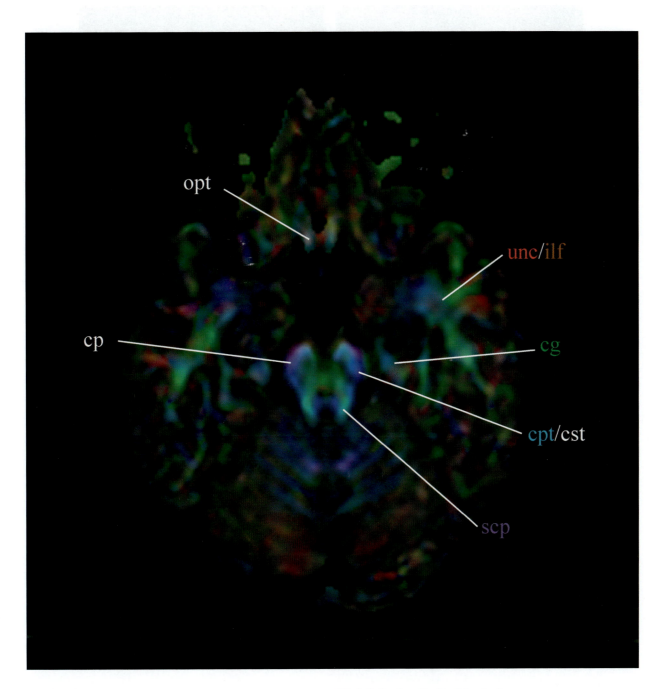

cg: cingulum
cp: cerebral peduncle
cpt: corticopontine tract
cst: corticospinal tract

ilf: inferior longitudinal fasciculus
opt: optic tract
scp: superior cerebellar peduncle
unc: uncinate fasciculus

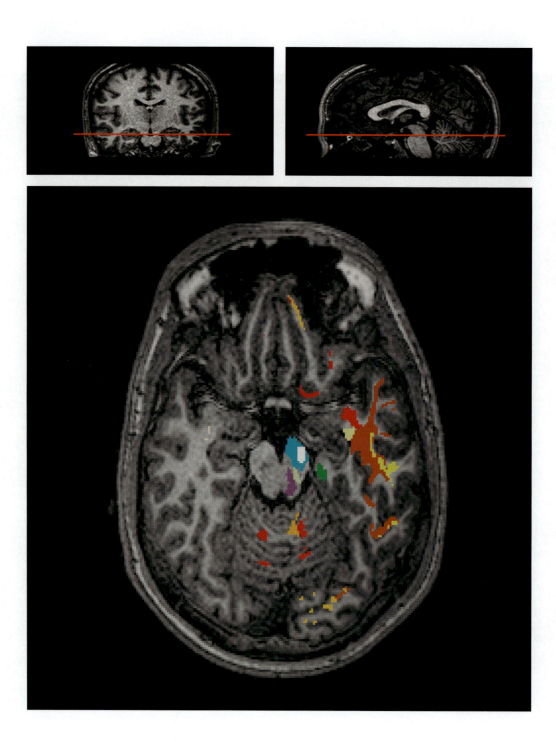

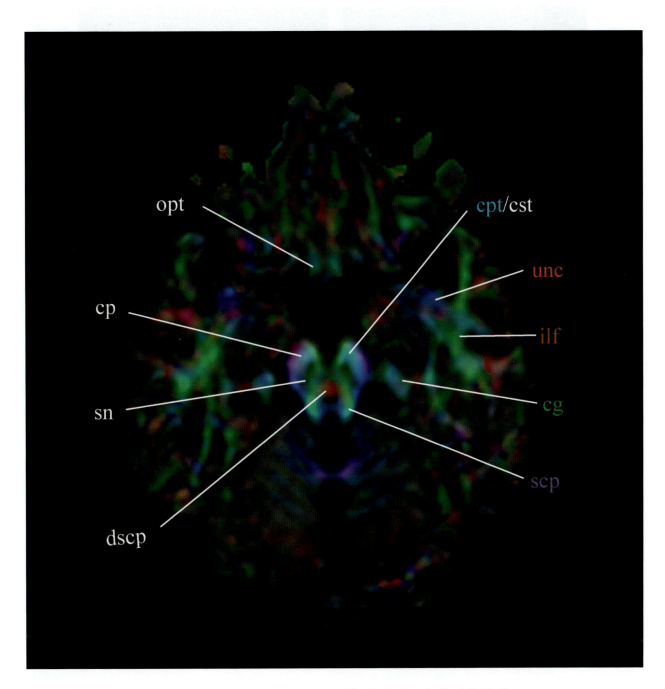

cg: cingulum
cp: cerebral peduncle
cpt: corticopontine tract
cst: corticospinal tract
dscp: decussation of the superior cerebellar peduncles

ilf: inferior longitudinal fasciculus
opt: optic tract
scp: superior cerebellar peduncle
sn: substantia nigra
unc: uncinate fasciculus

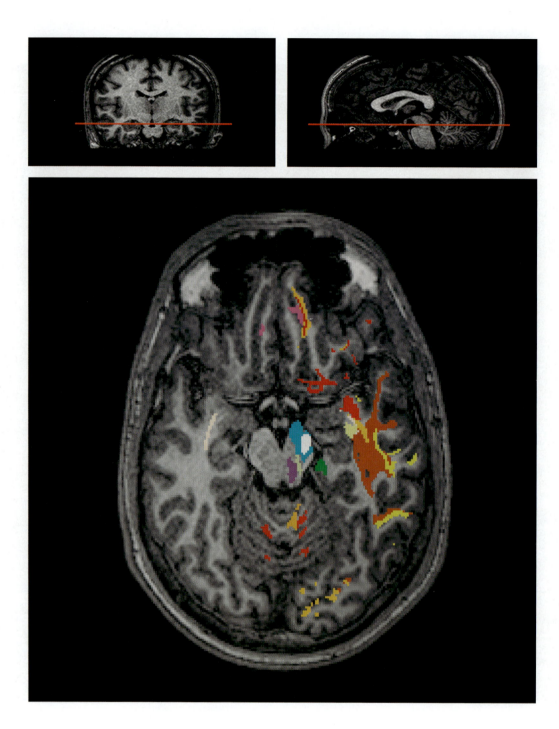

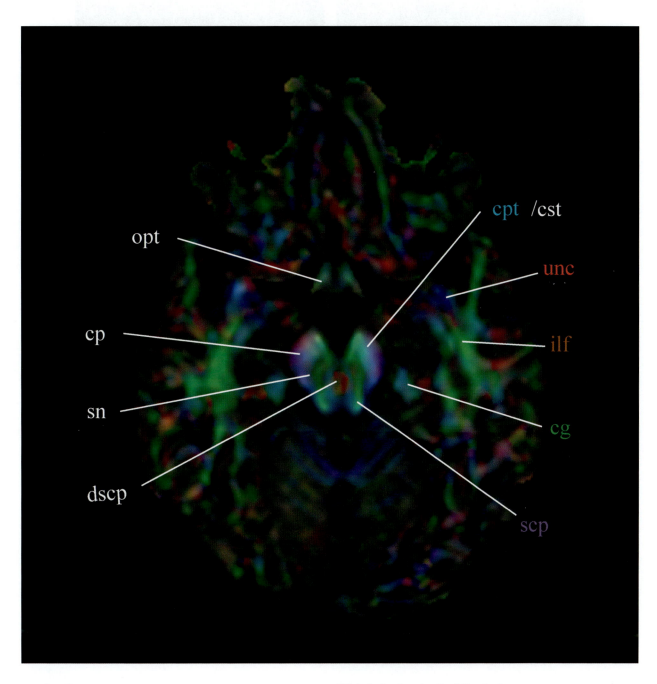

cg: cingulum
cp: cerebral peduncle
cpt: corticopontine tract
cst: corticospinal tract
dscp: decussation of the superior cerebellar peduncles

ilf: inferior longitudinal fasciculus
opt: optic tract
scp: superior cerebellar peduncle
sn: substantia nigra
unc: uncinate fasciculus

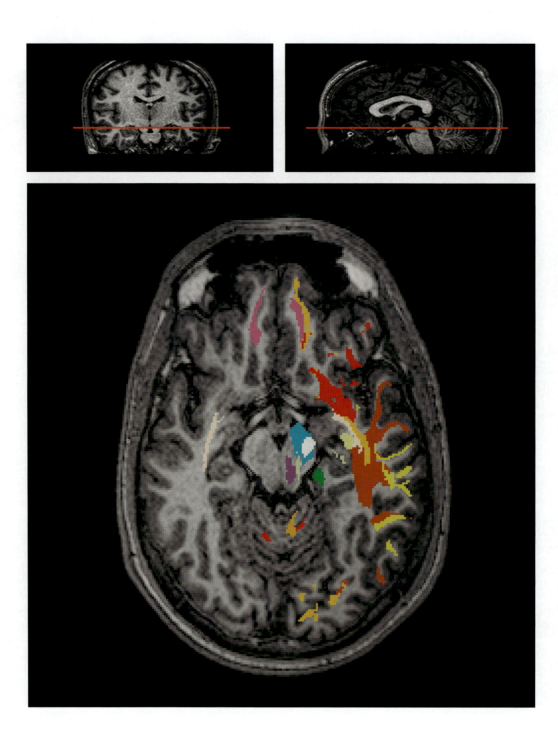

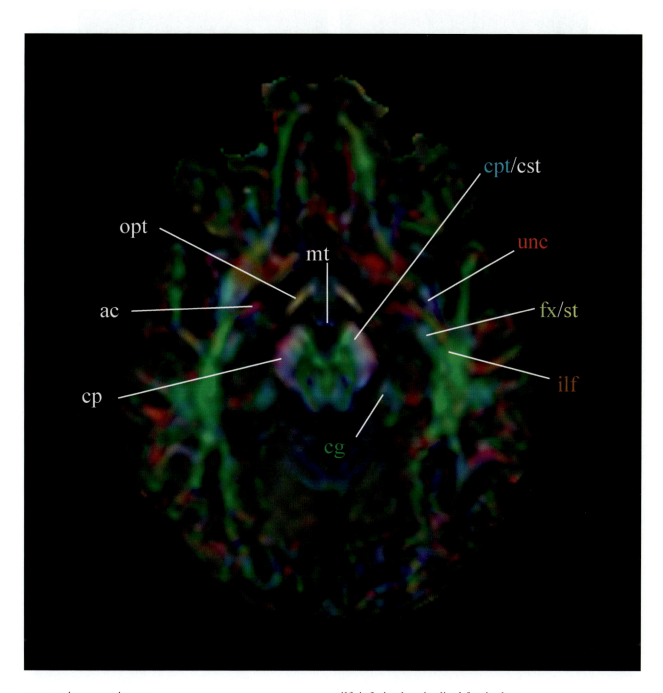

ac: anterior commissure
cg: cingulum
cp: cerebral peduncle
cpt: corticopontine tract
cst: corticospinal tract
fx: fornix

ilf: inferior longitudinal fasciculus
mt: mammillothalamic tract
opt: optic tract
st: stria terminalis
unc: uncinate fasciculus

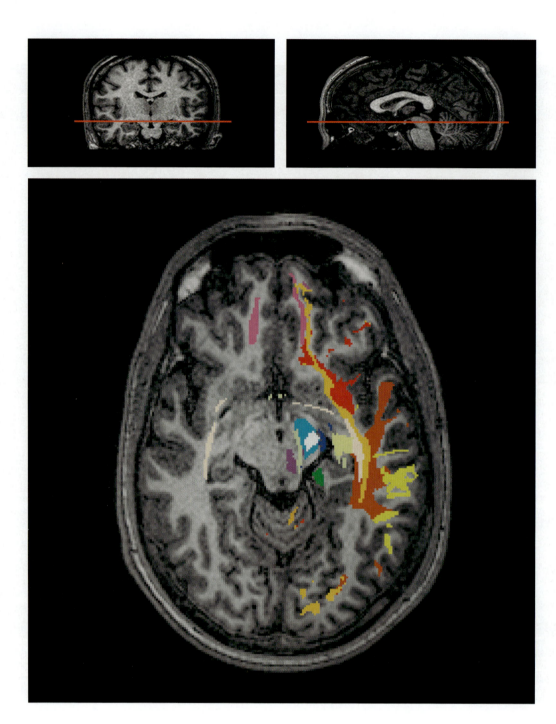

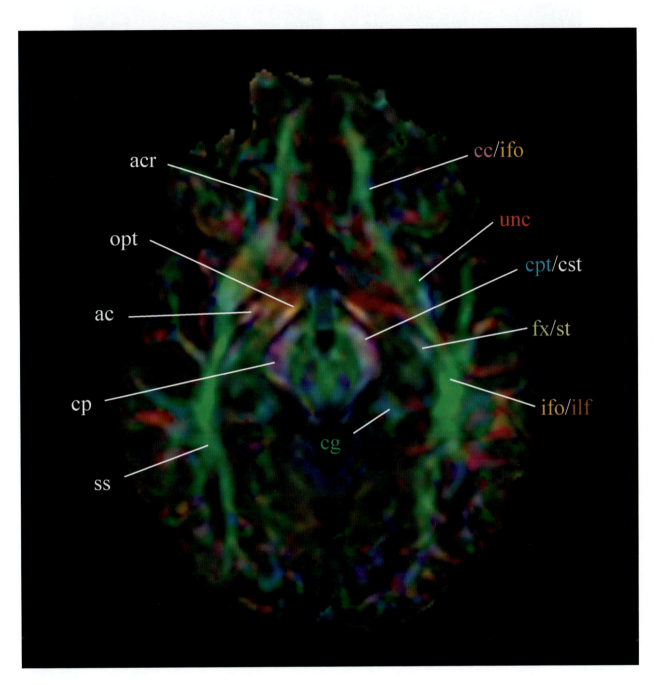

ac: anterior commissure
acr: anterior corona radiata
cc: corpus callosum
cg: cingulum
cp: cerebral peduncle
cpt: corticopontine tract
cst: corticospinal tract

fx: fornix
ifo: inferior fronto-occipital fasciculus
ilf: inferior longitudinal fasciculus
opt: optic tract
ss: sagittal stratum
st: stria terminalis
unc: uncinate fasciculus

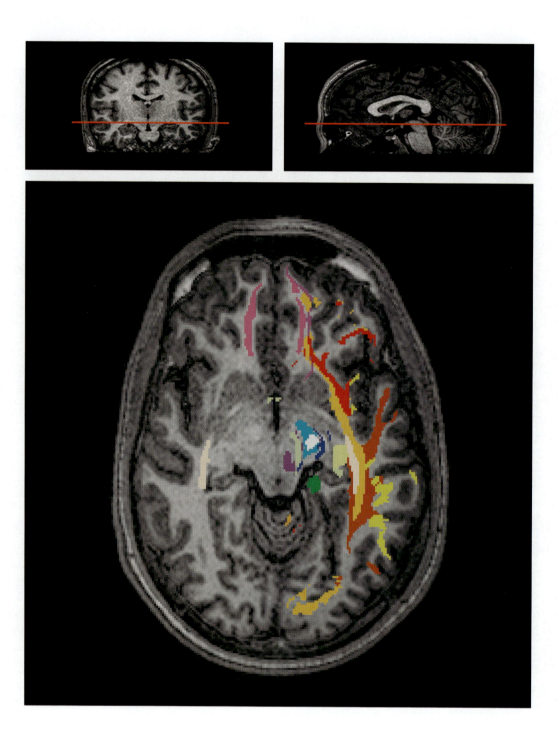

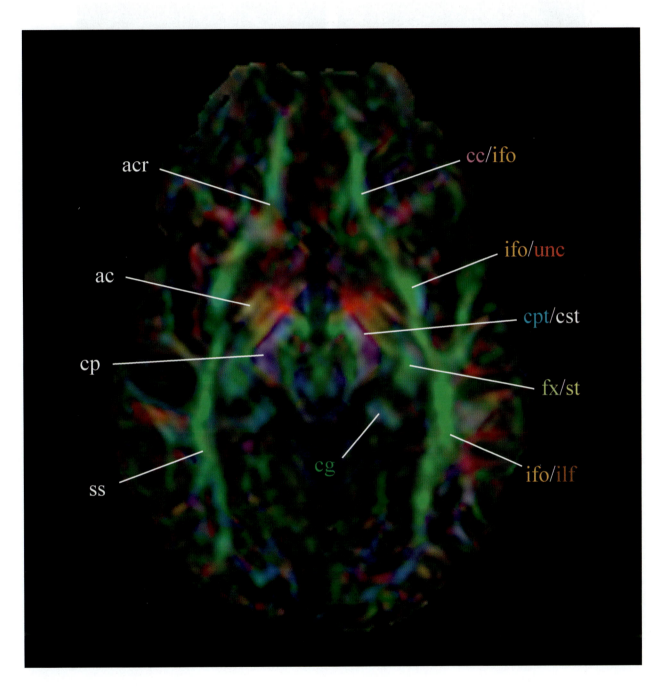

ac: anterior commissure
acr: anterior corona radiata
cc: corpus callosum
cg: cingulum
cp: cerebral peduncle
cpt: corticopontine tract
cst: corticospinal tract

fx: fornix
ifo: inferior fronto-occipital fasciculus
ilf: inferior longitudinal fasciculus
ss: sagittal stratum
st: stria terminalis
unc: uncinate fasciculus

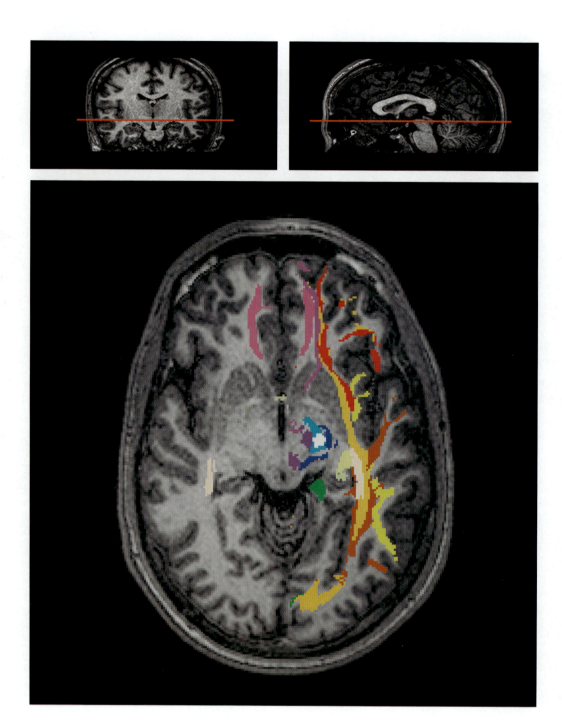

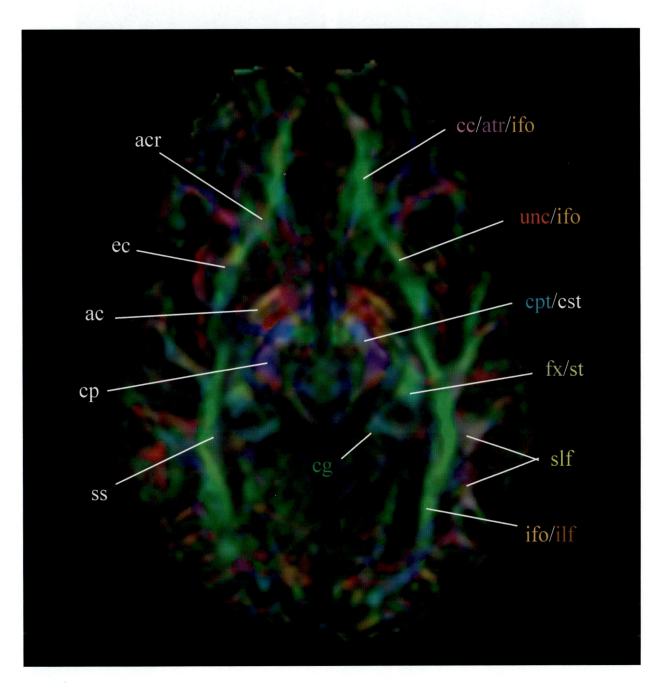

ac: anterior commissure
acr: anterior corona radiata
atr: anterior thalamic radiation
cc: corpus callosum
cg: cingulum
cp: cerebral peduncle
cpt: corticopontine tract
cst: corticospinal tract
ec: external capsule

fx: fornix
ifo: inferior fronto-occipital fasciculus
ilf: inferior longitudinal fasciculus
slf: superior longitudinal fasciculus
ss: sagittal stratum
st: stria terminalis
unc: uncinate fasciculus

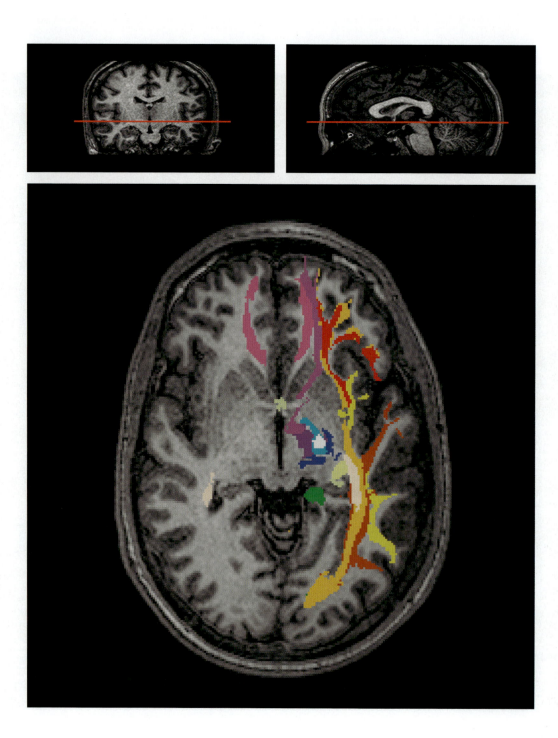

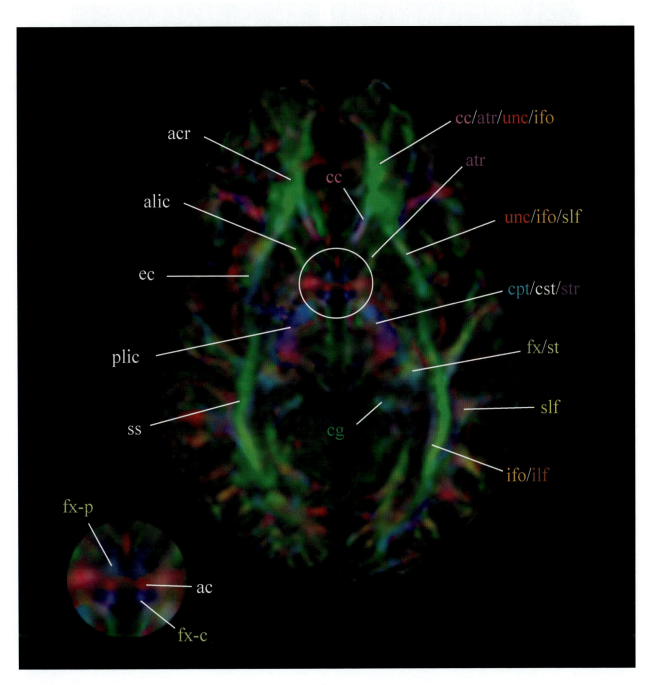

ac: anterior commissure
acr: anterior corona radiata
alic: anterior limb of the internal capsule
atr: anterior thalamic radiation
cc: corpus callosum
cg: cingulum
cpt: corticopontine tract
cst: corticospinal tract
ec: external capsule
fx: fornix

fx-c: column of the fornix
fx-p: precommissural part of the fornix
ifo: inferior fronto-occipital fasciculus
ilf: inferior longitudinal fasciculus
plic: posterior limb of the internal capsule
slf: superior longitudinal fasciculus
ss: sagittal stratum
st: stria terminalis
str: superior thalamic radiation
unc: uncinate fasciculus

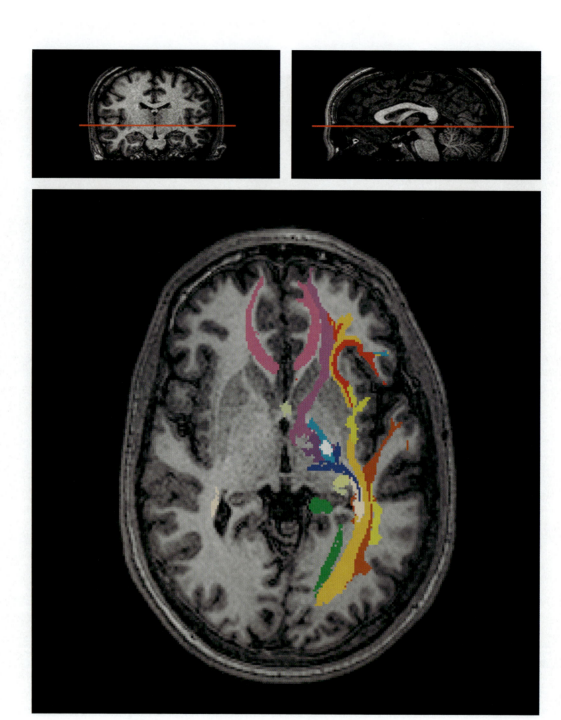

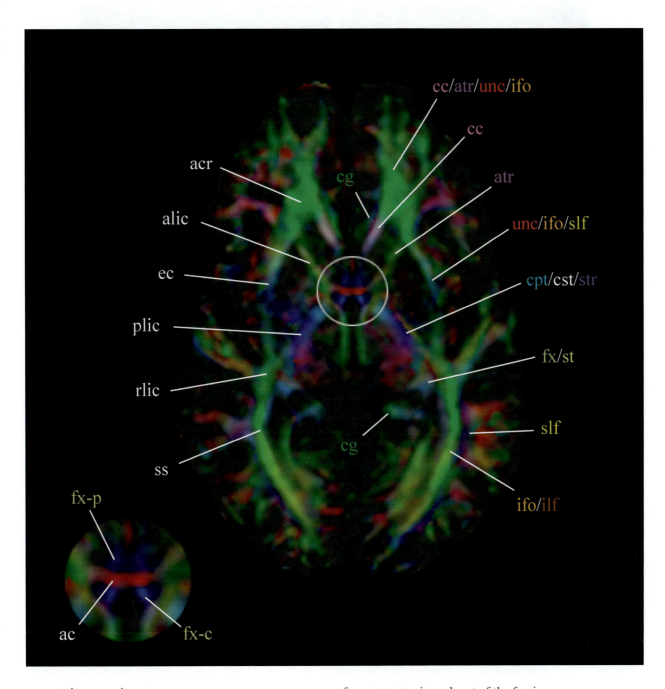

ac: anterior commissure
acr: anterior corona radiata
alic: anterior limb of the internal capsule
atr: anterior thalamic radiation
cc: corpus callosum
cg: cingulum
cpt: corticopontine tract
cst: corticospinal tract
ec: external capsule
fx: fornix
fx-c: column of the fornix

fx-p: precommissural part of the fornix
ifo: inferior fronto-occipital fasciculus
ilf: inferior longitudinal fasciculus
plic: posterior limb of the internal capsule
rlic: retrolenticular part of the internal capsule
slf: superior longitudinal fasciculus
ss: sagittal stratum
st: stria terminalis
str: superior thalamic radiation
unc: uncinate fasciculus

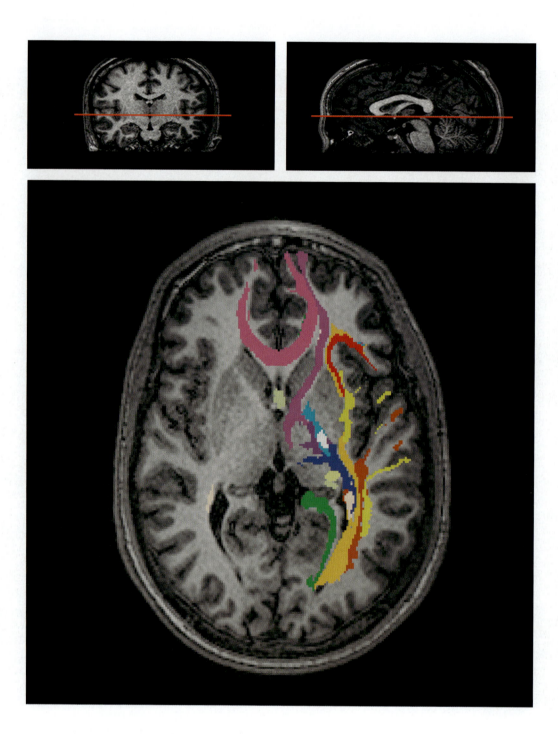

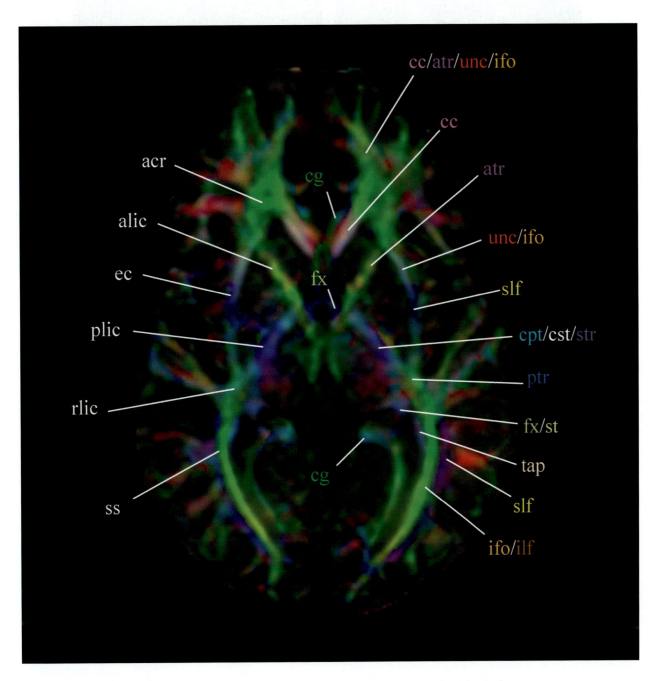

acr: anterior corona radiata
alic: anterior limb of the internal capsule
atr: anterior thalamic radiation
cc: corpus callosum
cg: cingulum
cpt: corticopontine tract
cst: corticospinal tract
ec: external capsule
fx: fornix
ifo: inferior fronto-occipital fasciculus

ilf: inferior longitudinal fasciculus
plic: posterior limb of the internal capsule
ptr: posterior thalamic radiation
rlic: retrolenticular part of the internal capsule
slf: superior longitudinal fasciculus
ss: sagittal stratum
st: stria terminalis
str: superior thalamic radiation
tap: tapetum
unc: uncinate fasciculus

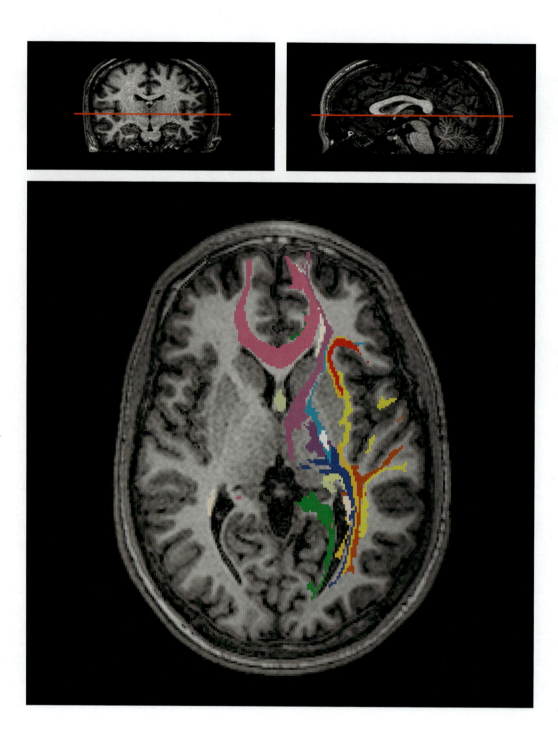

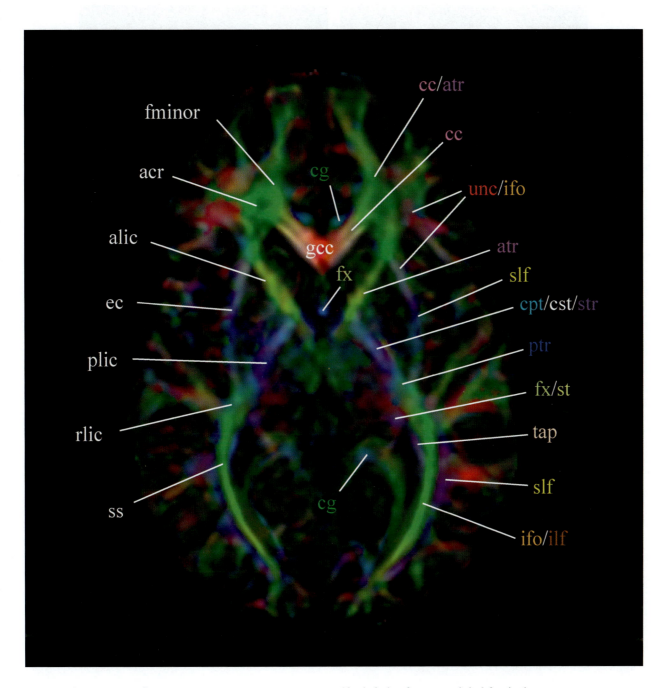

acr: anterior corona radiata
alic: anterior limb of the internal capsule
atr: anterior thalamic radiation
cc: corpus callosum
cg: cingulum
cpt: corticopontine tract
cst: corticospinal tract
ec: external capsule
fminor: forceps minor
fx: fornix
gcc: genu of the corpus callosum

ifo: inferior fronto-occipital fasciculus
ilf: inferior longitudinal fasciculus
plic: posterior limb of the internal capsule
ptr: posterior thalamic radiation
rlic: retrolenticular part of the internal capsule
slf: superior longitudinal fasciculus
ss: sagittal stratum
st: stria terminalis
str: superior thalamic radiation
tap: tapetum
unc: uncinate fasciculus

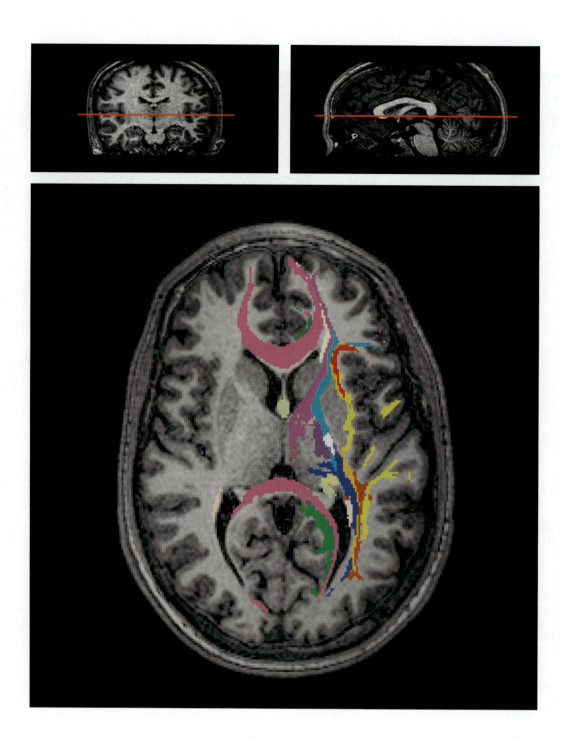

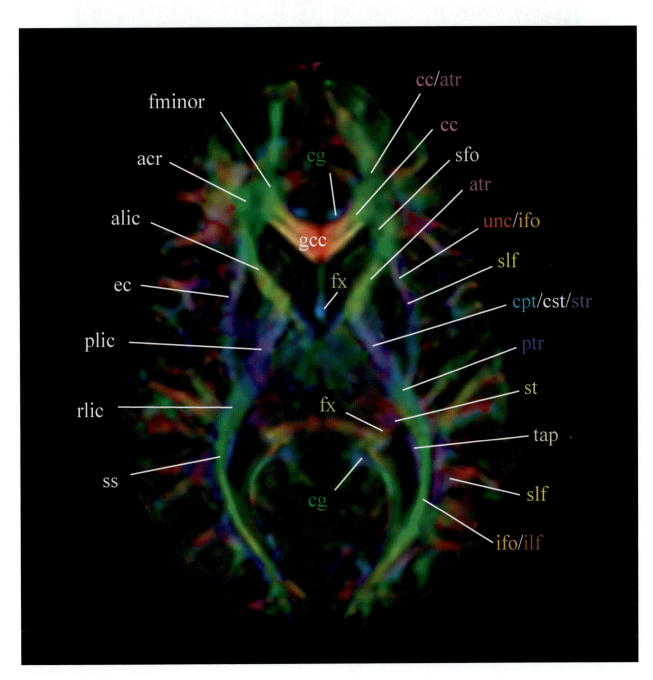

acr: anterior corona radiata
alic: anterior limb of the internal capsule
atr: anterior thalamic radiation
cc: corpus callosum
cg: cingulum
cpt: corticopontine tract
cst: corticospinal tract
ec: external capsule
fminor: forceps minor
fx: fornix
gcc: genu of the corpus callosum
ifo: inferior fronto-occipital fasciculus

ilf: inferior longitudinal fasciculus
plic: posterior limb of the internal capsule
ptr: posterior thalamic radiation
rlic: retrolenticular part of the internal capsule
sfo: superior fronto-occipital fasciculus
slf: superior longitudinal fasciculus
ss: sagittal stratum
st: stria terminalis
str: superior thalamic radiation
tap: tapetum
unc: uncinate fasciculus

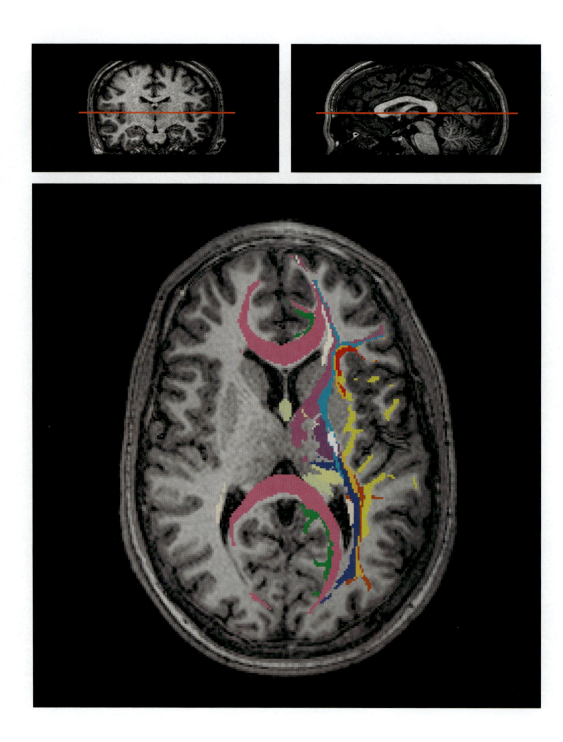

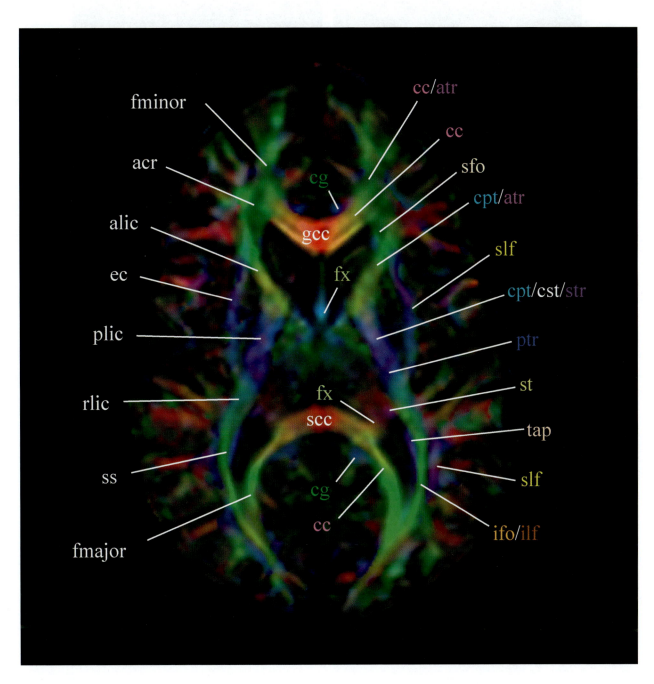

acr: anterior corona radiata
alic: anterior limb of the internal capsule
atr: anterior thalamic radiation
cc: corpus callosum
cg: cingulum
cpt: corticopontine tract
cst: corticospinal tract
ec: external capsule
fmajor: forceps major
fminor: forceps minor
fx: fornix
gcc: genu of the corpus callosum

ifo: inferior fronto-occipital fasciculus
ilf: inferior longitudinal fasciculus
plic: posterior limb of the internal capsule
ptr: posterior thalamic radiation
rlic: retrolenticular part of the internal capsule
scc: splenium of the corpus callosum
sfo: superior fronto-occipital fasciculus
slf: superior longitudinal fasciculus
ss: sagittal stratum
st: stria terminalis
str: superior thalamic radiation
tap: tapetum

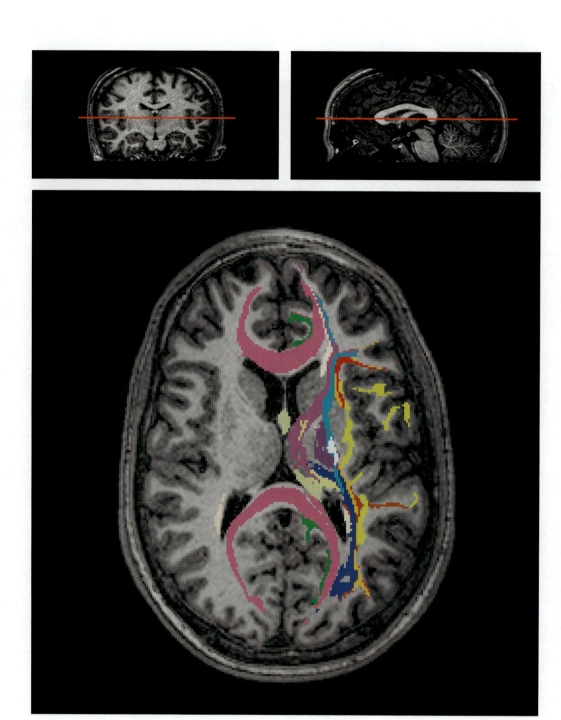

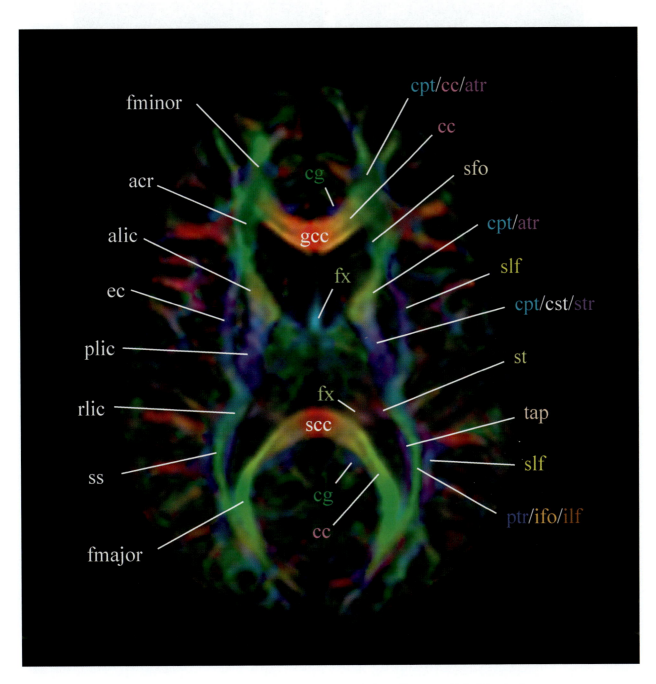

acr: anterior corona radiata
alic: anterior limb of the internal capsule
atr: anterior thalamic radiation
cc: corpus callosum
cg: cingulum
cpt: corticopontine tract
cst: corticospinal tract
ec: external capsule
fmajor: forceps major
fminor: forceps minor
fx: fornix
gcc: genu of the corpus callosum

ifo: inferior fronto-occipital fasciculus
ilf: inferior longitudinal fasciculus
plic: posterior limb of the internal capsule
ptr: posterior thalamic radiation
rlic: retrolenticular part of the internal capsule
scc: splenium of the corpus callosum
sfo: superior fronto-occipital fasciculus
slf: superior longitudinal fasciculus
ss: sagittal stratum
st: stria terminalis
str: superior thalamic radiation
tap: tapetum

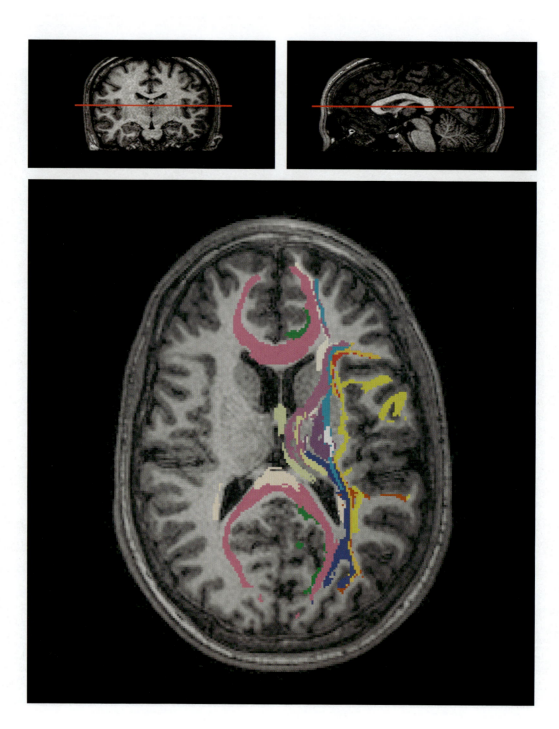

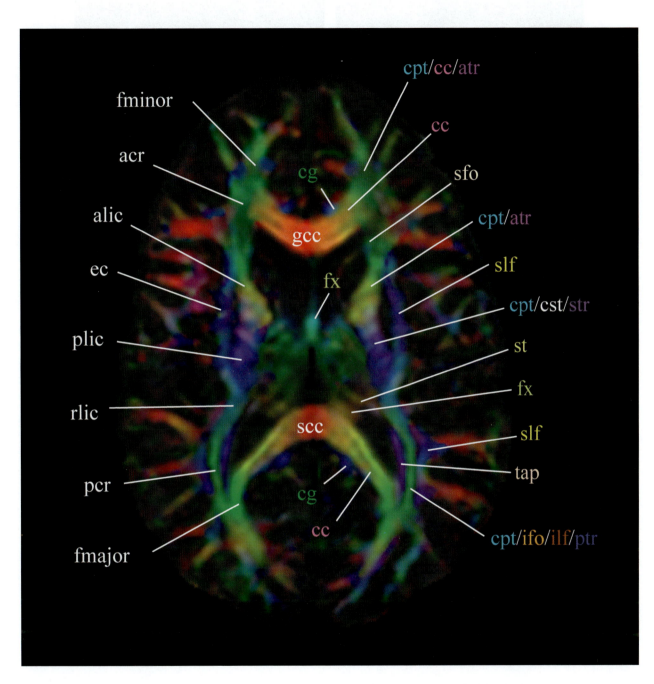

acr: anterior corona radiata
alic: anterior limb of the internal capsule
atr: anterior thalamic radiation
cc: corpus callosum
cg: cingulum
cpt: corticopontine tract
cst: corticospinal tract
ec: external capsule
fmajor: forceps major
fminor: forceps minor
fx: fornix
gcc: genu of the corpus callosum

ifo: inferior fronto-occipital fasciculus
ilf: inferior longitudinal fasciculus
pcr: posterior corona radiata
plic: posterior limb of the internal capsule
ptr: posterior thalamic radiation
rlic: retrolenticular part of the internal capsule
scc: splenium of the corpus callosum
sfo: superior fronto-occipital fasciculus
slf: superior longitudinal fasciculus
st: stria terminalis
str: superior thalamic radiation
tap: tapetum

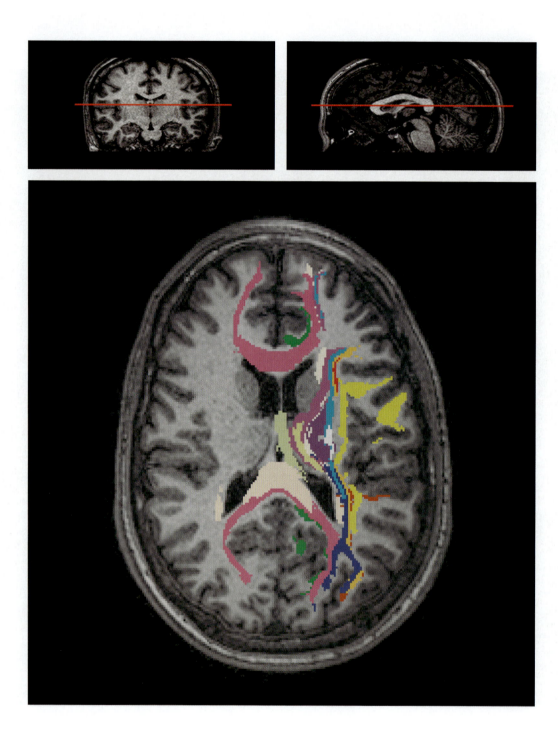

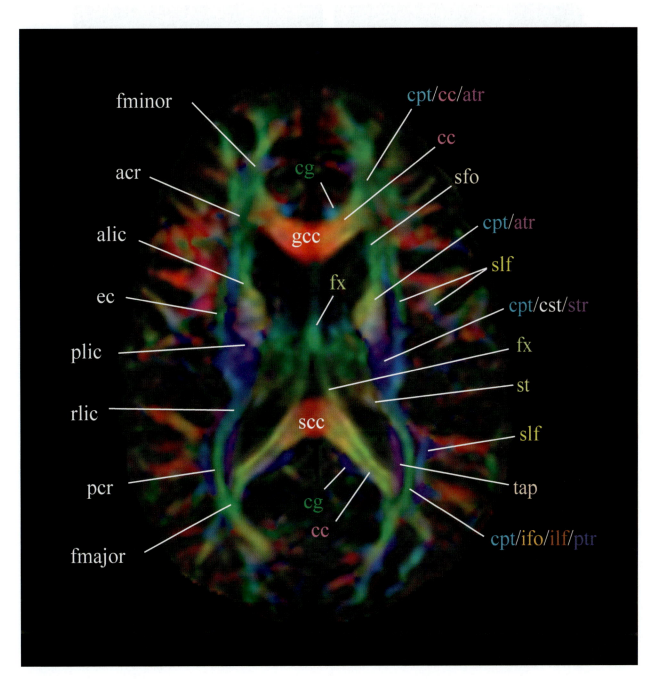

acr: anterior corona radiata
alic: anterior limb of the internal capsule
atr: anterior thalamic radiation
cc: corpus callosum
cg: cingulum
cpt: corticopontine tract
cst: corticospinal tract
ec: external capsule
fmajor: forceps major
fminor: forceps minor
fx: fornix
gcc: genu of the corpus callosum

ifo: inferior fronto-occipital fasciculus
ilf: inferior longitudinal fasciculus
pcr: posterior corona radiata
plic: posterior limb of the internal capsule
ptr: posterior thalamic radiation
rlic: retrolenticular part of the internal capsule
scc: splenium of the corpus callosum
sfo: superior fronto-occipital fasciculus
slf: superior longitudinal fasciculus
st: stria terminalis
str: superior thalamic radiation
tap: tapetum

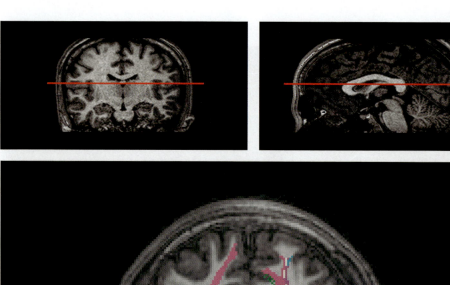

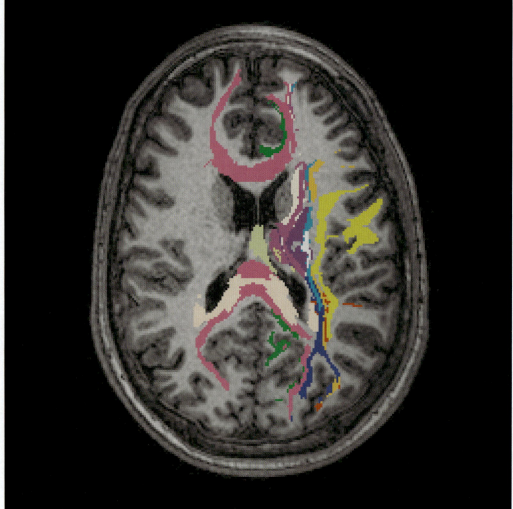

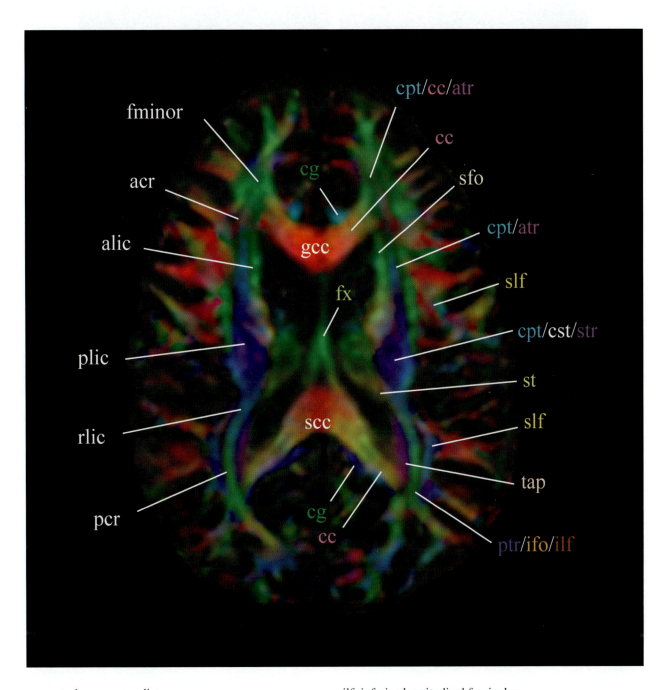

acr: anterior corona radiata
alic: anterior limb of the internal capsule
atr: anterior thalamic radiation
cc: corpus callosum
cg: cingulum
cpt: corticopontine tract
cst: corticospinal tract
fminor: forceps minor
fx: fornix
gcc: genu of the corpus callosum
ifo: inferior fronto-occipital fasciculus

ilf: inferior longitudinal fasciculus
pcr: posterior corona radiata
plic: posterior limb of the internal capsule
ptr: posterior thalamic radiation
rlic: retrolenticular part of the internal capsule
scc: splenium of the corpus callosum
sfo: superior fronto-occipital fasciculus
slf: superior longitudinal fasciculus
st: stria terminalis
str: superior thalamic radiation
tap: tapetum

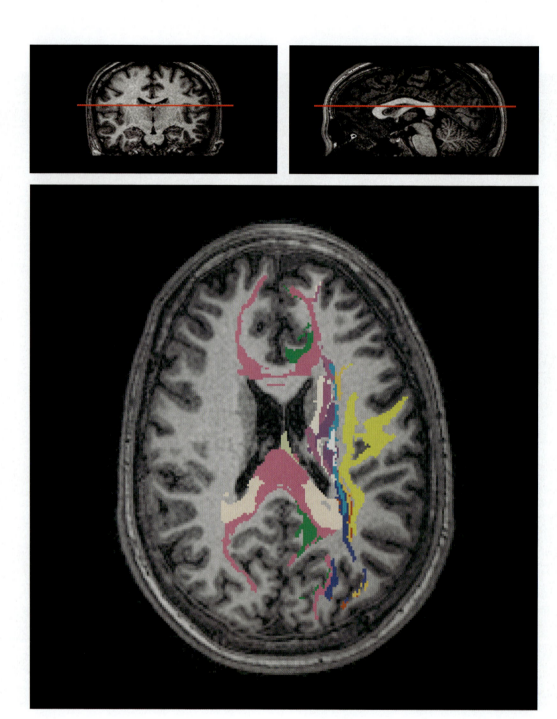

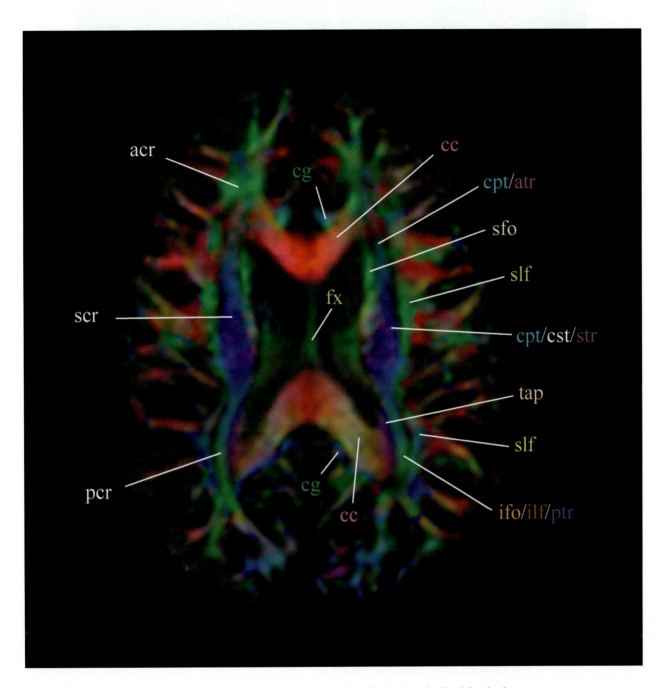

acr: anterior corona radiata
atr: anterior thalamic radiation
cc: corpus callosum
cg: cingulum
cpt: corticopontine tract
cst: corticospinal tract
fx: fornix
ifo: inferior fronto-occipital fasciculus

ilf: inferior longitudinal fasciculus
pcr: posterior corona radiata
ptr: posterior thalamic radiation
scr: superior corona radiata
sfo: superior fronto-occipital fasciculus
slf: superior longitudinal fasciculus
str: superior thalamic radiation
tap: tapetum

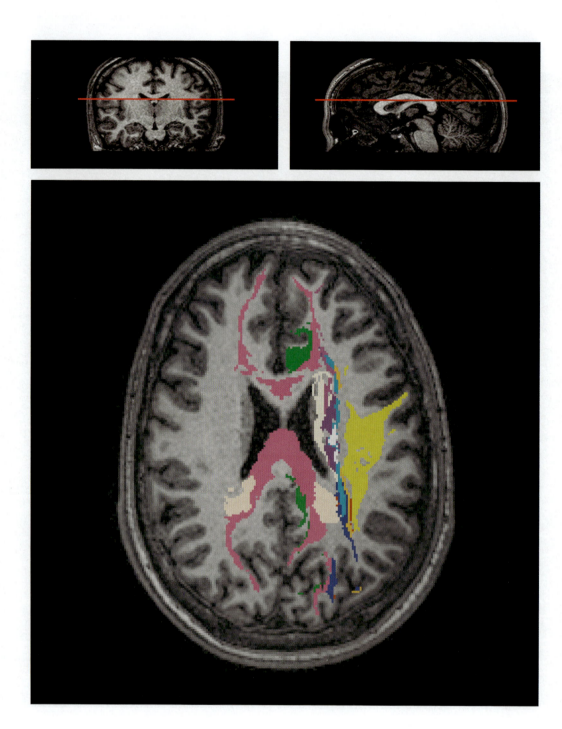

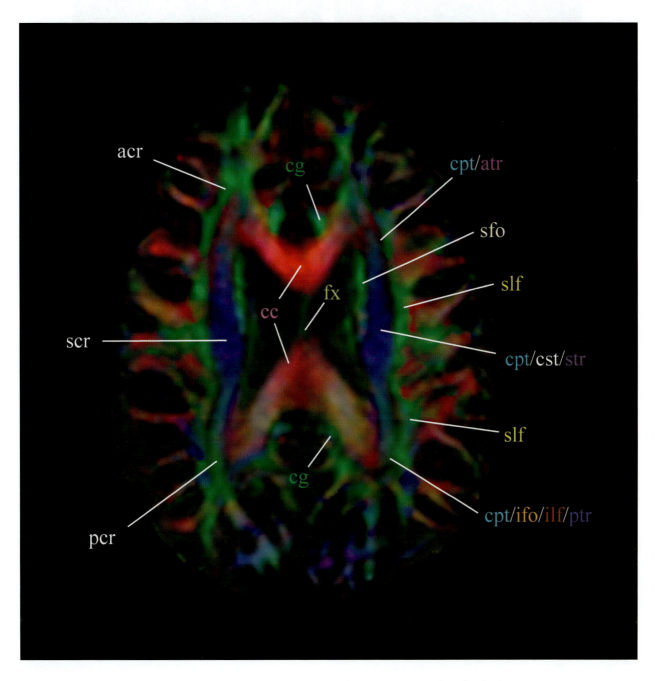

acr: anterior corona radiata
atr: anterior thalamic radiation
cc: corpus callosum
cg: cingulum
cpt: corticopontine tract
cst: corticospinal tract
fx: fornix
ifo: inferior fronto-occipital fasciculus

ilf: inferior longitudinal fasciculus
pcr: posterior corona radiata
ptr: posterior thalamic radiation
scr: superior corona radiata
sfo: superior fronto-occipital fasciculus
slf: superior longitudinal fasciculus
str: superior thalamic radiation

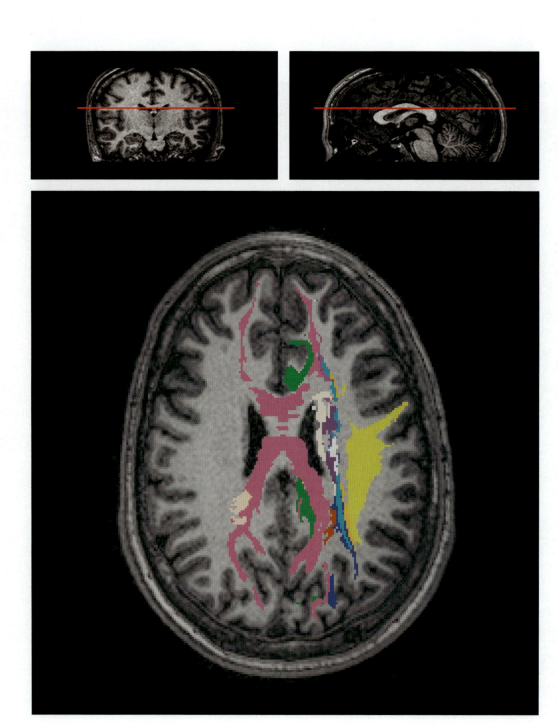

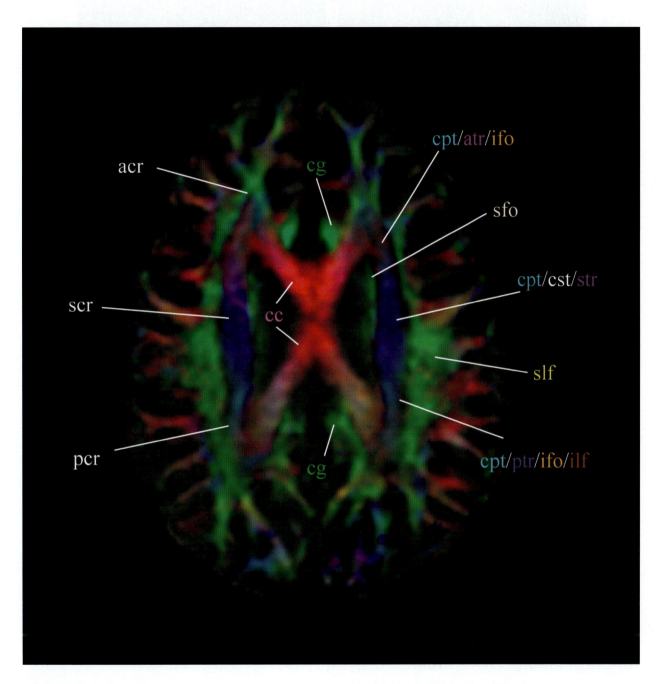

acr: anterior corona radiata
atr: anterior thalamic radiation
cc: corpus callosum
cg: cingulum
cpt: corticopontine tract
cst: corticospinal tract
ifo: inferior fronto-occipital fasciculus

ilf: inferior longitudinal fasciculus
pcr: posterior corona radiata
ptr: posterior thalamic radiation
scr: superior corona radiata
sfo: superior fronto-occipital fasciculus
slf: superior longitudinal fasciculus
str: superior thalamic radiation

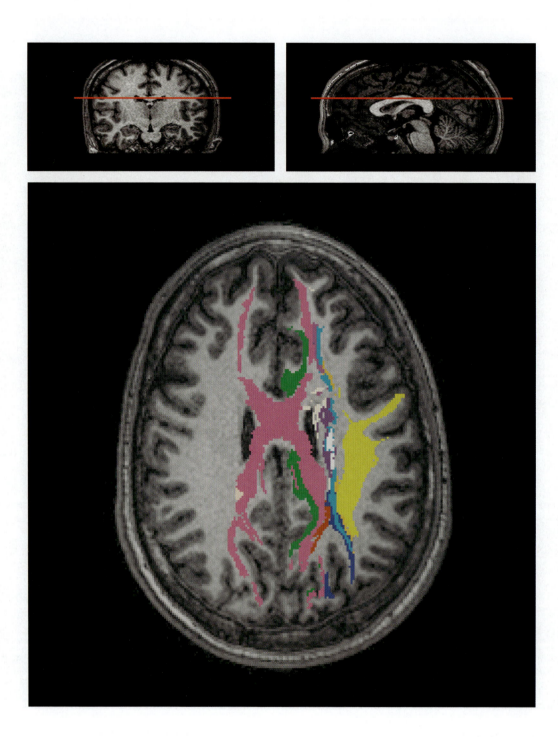

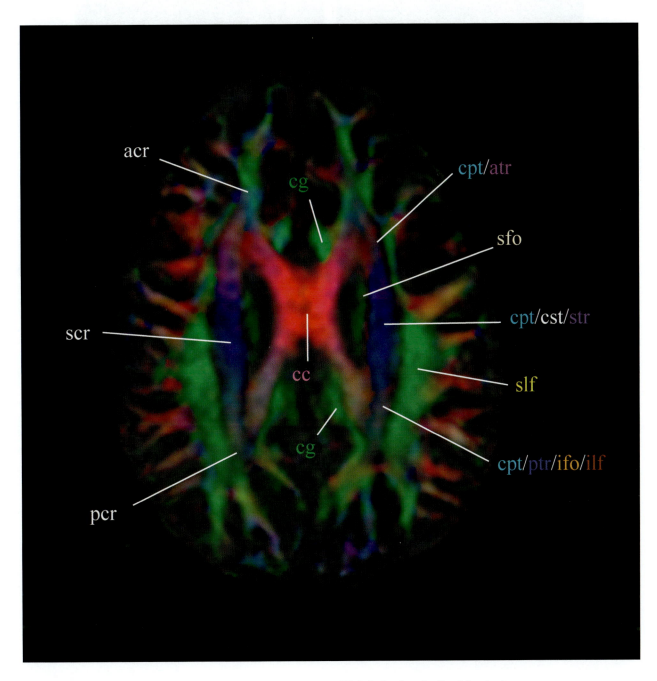

acr: anterior corona radiata
atr: anterior thalamic radiation
cc: corpus callosum
cg: cingulum
cpt: corticopontine tract
cst: corticospinal tract
ifo: inferior fronto-occipital fasciculus

ilf: inferior longitudinal fasciculus
pcr: posterior corona radiata
ptr: posterior thalamic radiation
scr: superior corona radiata
sfo: superior fronto-occipital fasciculus
slf: superior longitudinal fasciculus
str: superior thalamic radiation

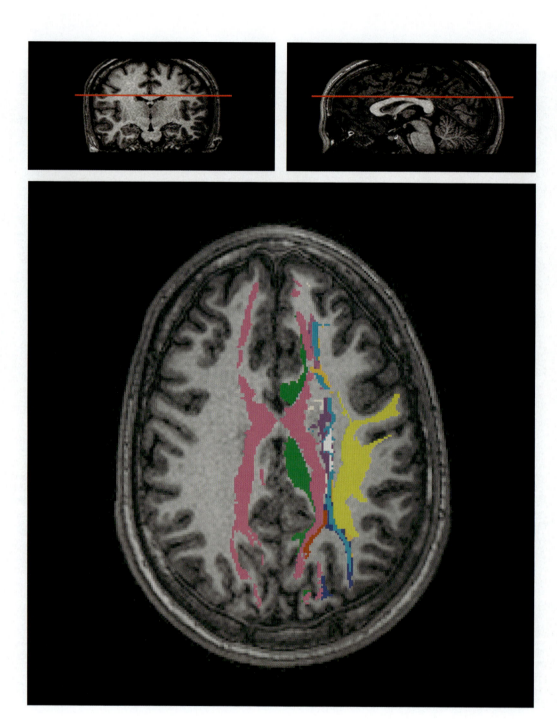

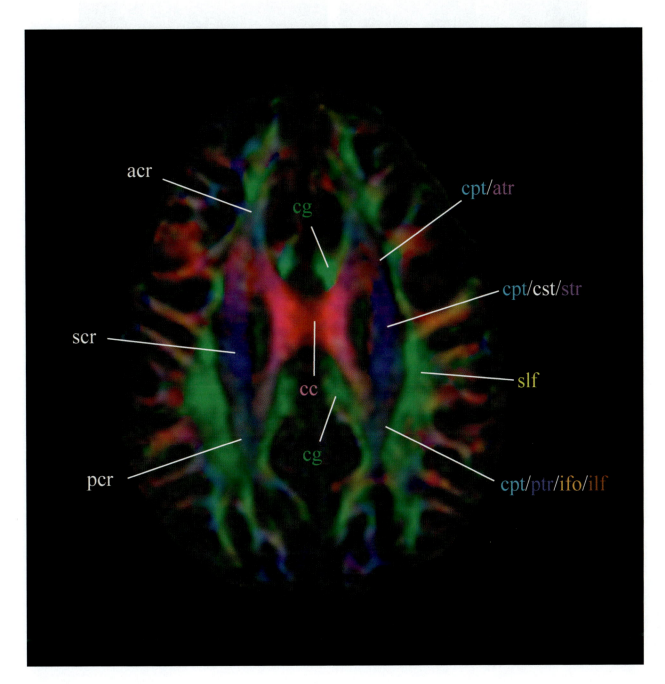

acr: anterior corona radiata
atr: anterior thalamic radiation
cc: corpus callosum
cg: cingulum
cpt: corticopontine tract
cst: corticospinal tract
ifo: inferior fronto-occipital fasciculus

ilf: inferior longitudinal fasciculus
pcr: posterior corona radiata
ptr: posterior thalamic radiation
scr: superior corona radiata
slf: superior longitudinal fasciculus
str: superior thalamic radiation

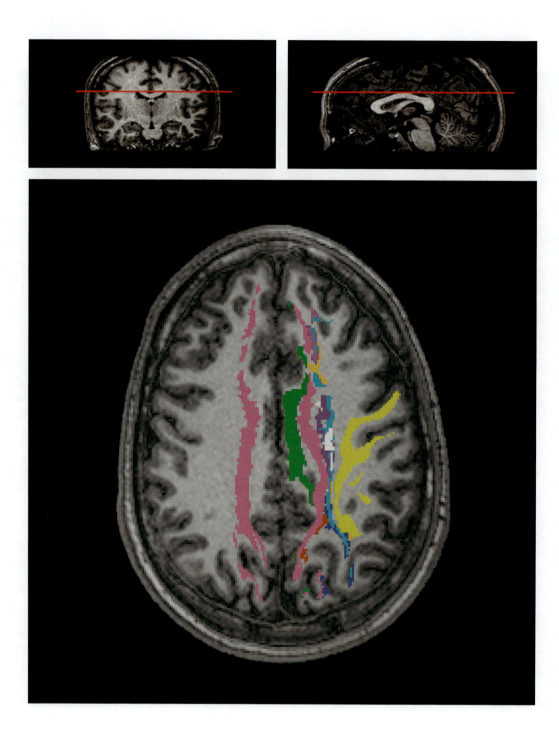

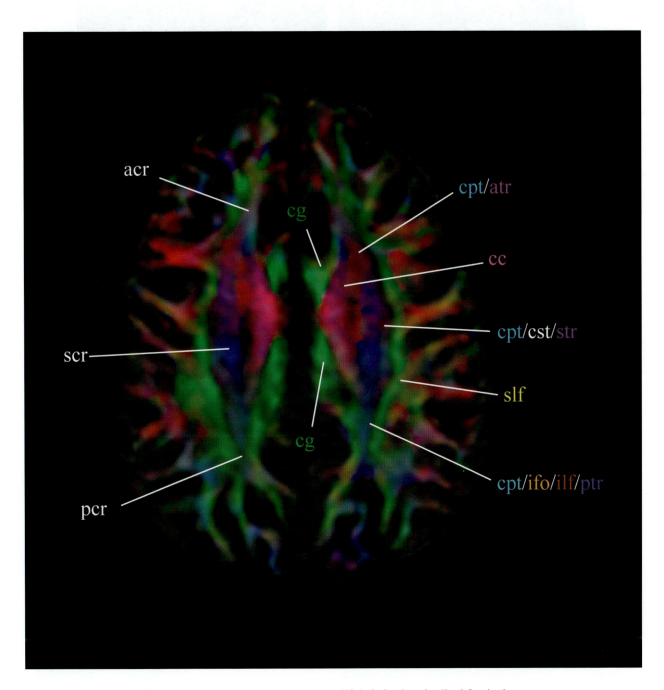

acr: anterior corona radiata
atr: anterior thalamic radiation
cc: corpus callosum
cg: cingulum
cpt: corticopontine tract
cst: corticospinal tract
ifo: inferior fronto-occipital fasciculus

ilf: inferior longitudinal fasciculus
pcr: posterior corona radiata
ptr: posterior thalamic radiation
scr: superior corona radiata
slf: superior longitudinal fasciculus
str: superior thalamic radiation

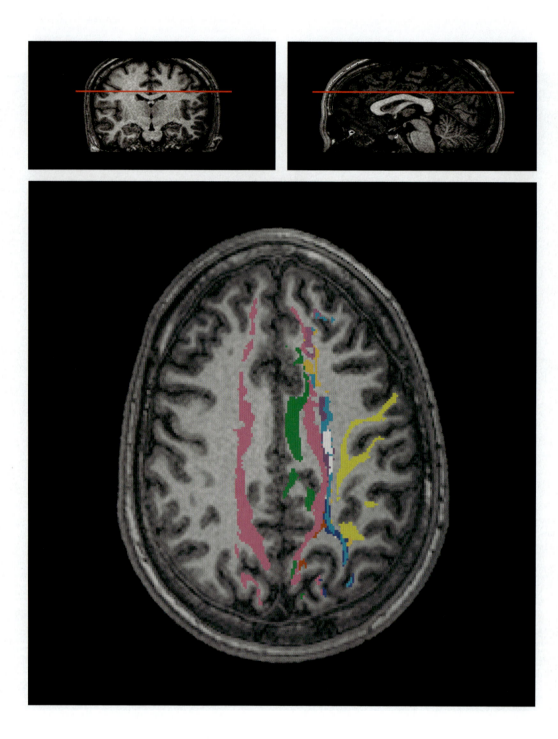

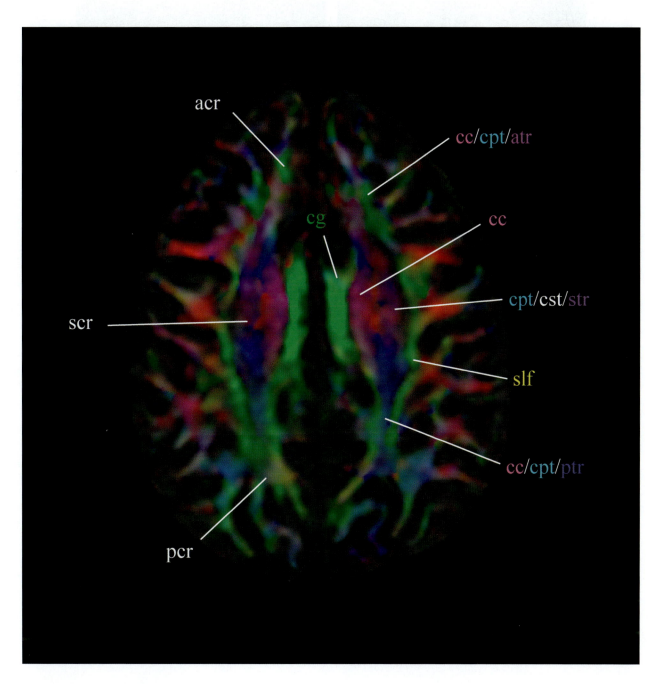

acr: anterior corona radiata
atr: anterior thalamic radiation
cc: corpus callosum
cg: cingulum
cpt: corticopontine tract
cst: corticospinal tract

pcr: posterior corona radiata
ptr: posterior thalamic radiation
scr: superior corona radiata
slf: superior longitudinal fasciculus
str: superior thalamic radiation

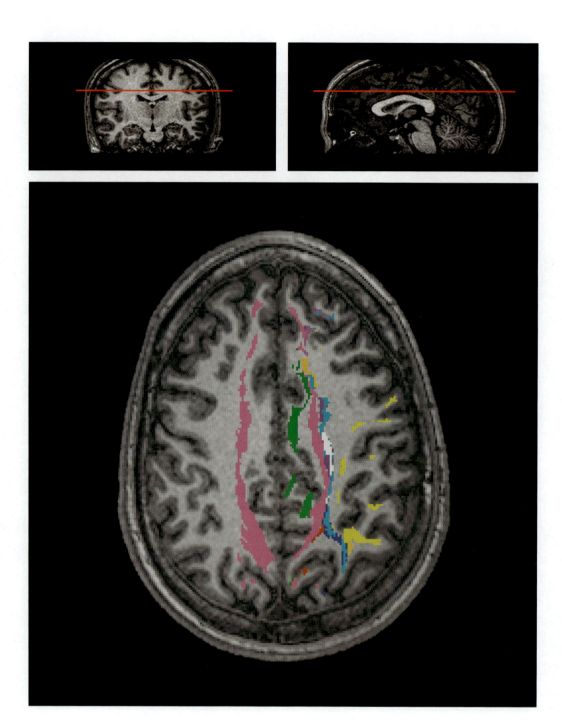

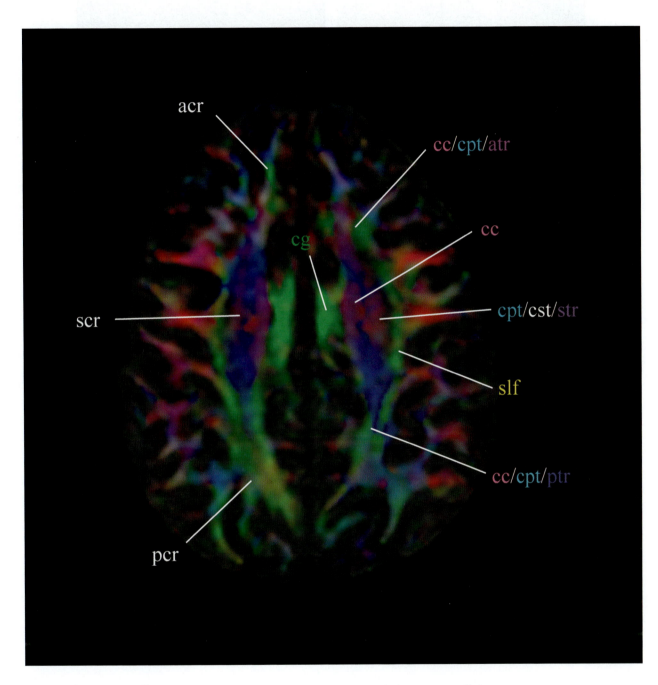

acr: anterior corona radiata
atr: anterior thalamic radiation
cc: corpus callosum
cg: cingulum
cpt: corticopontine tract
cst: corticospinal tract

pcr: posterior corona radiata
ptr: posterior thalamic radiation
scr: superior corona radiata
slf: superior longitudinal fasciculus
str: superior thalamic radiation

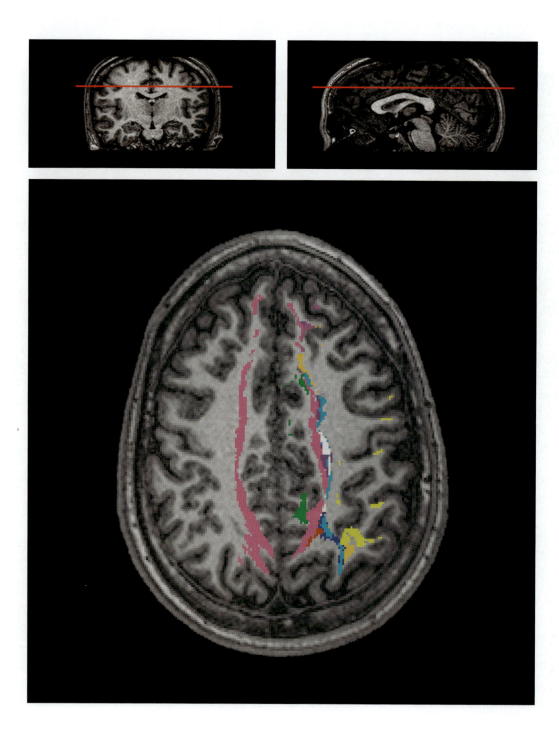

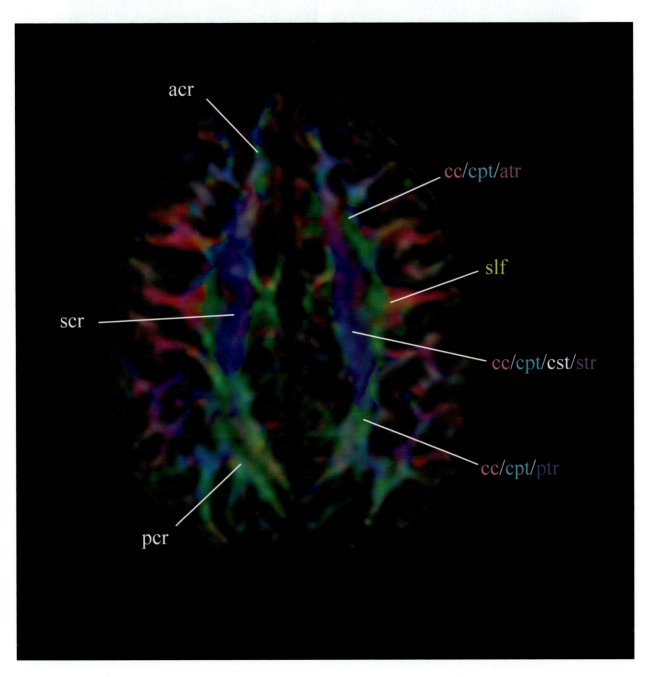

acr: anterior corona radiata
atr: anterior thalamic radiation
cc: corpus callosum
cpt: corticopontine tract
cst: corticospinal tract

pcr: posterior corona radiata
ptr: posterior thalamic radiation
scr: superior corona radiata
slf: superior longitudinal fasciculus
str: superior thalamic radiation

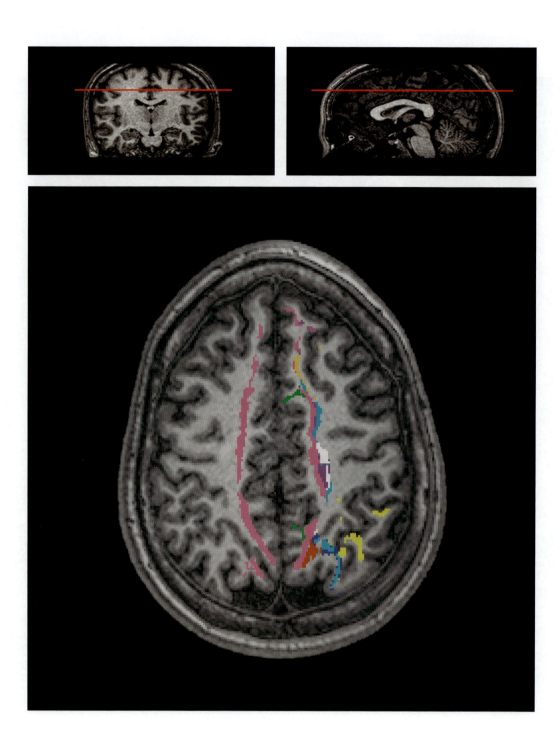

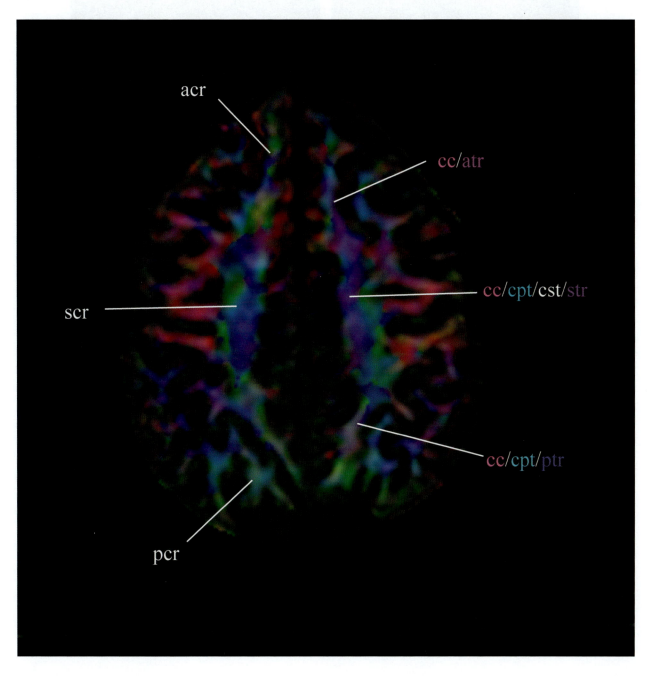

acr: anterior corona radiata
atr: anterior thalamic radiation
cc: corpus callosum
cpt: corticopontine tract
cst: corticospinal tract

pcr: posterior corona radiata
ptr: posterior thalamic radiation
scr: superior corona radiata
str: superior thalamic radiation

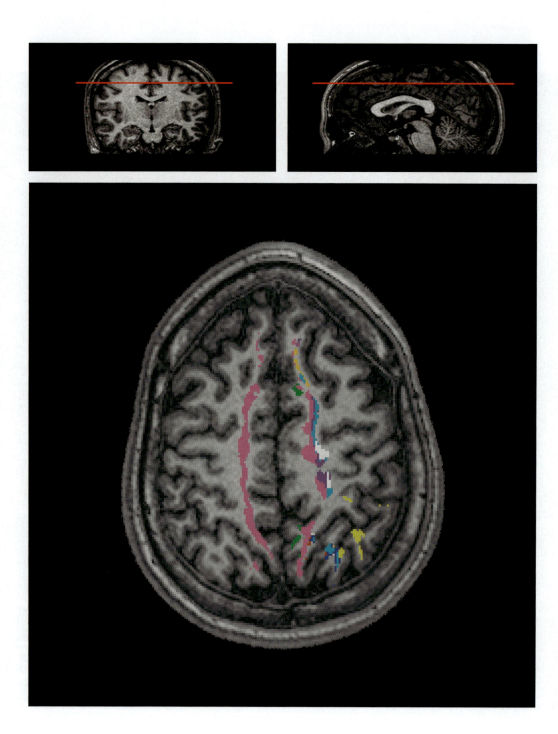

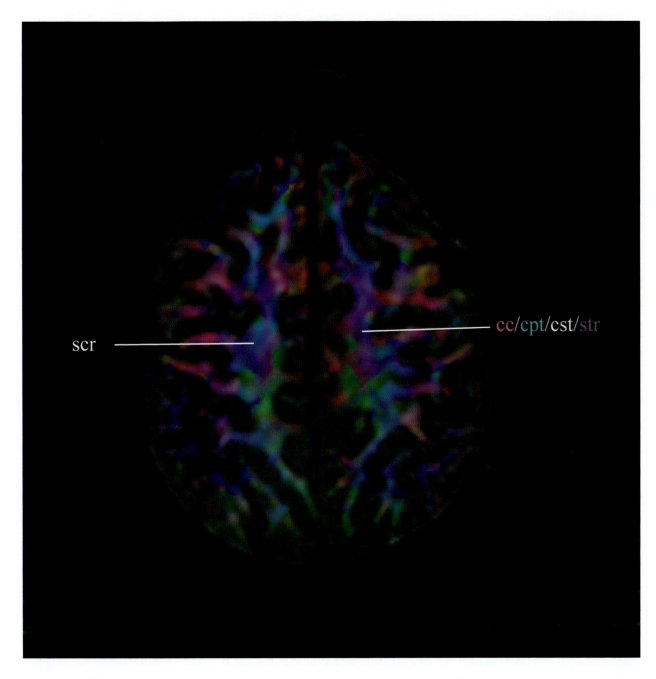

scr

cc/cpt/cst/str

cc: corpus callosum
cpt: corticopontine tract
cst: corticospinal tract

scr: superior corona radiata
str: superior thalamic radiation

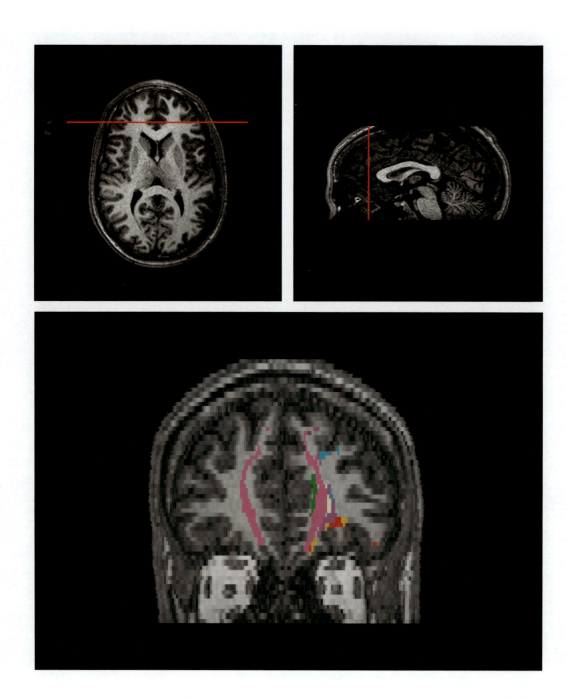

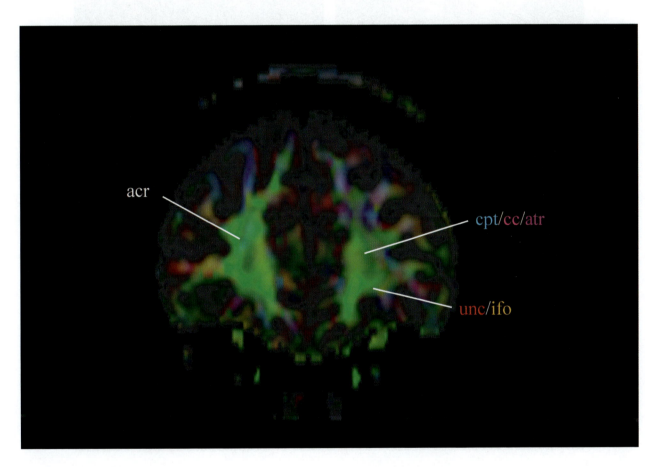

acr: anterior corona radiata
atr: anterior thalamic radiation
cc: corpus callosum

cpt: corticopontine tract
ifo: inferior fronto-occipital fasciculus
unc: uncinate fasciculus

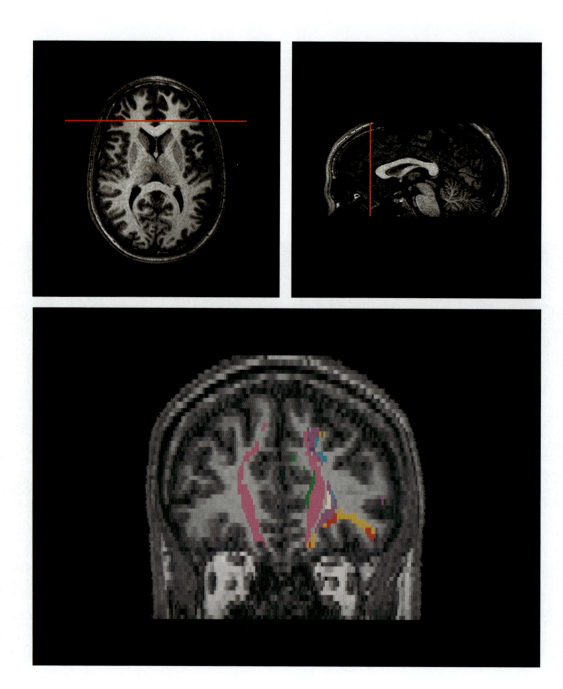

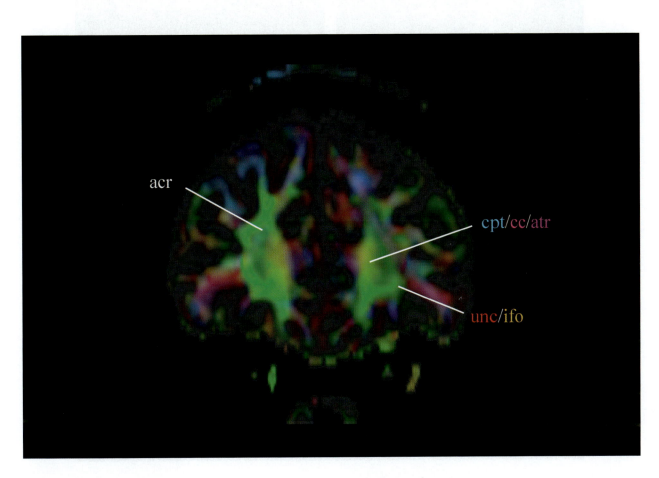

acr: anterior corona radiata
atr: anterior thalamic radiation
cc: corpus callosum

cpt: corticopontine tract
ifo: inferior fronto-occipital fasciculus
unc: uncinate fasciculus

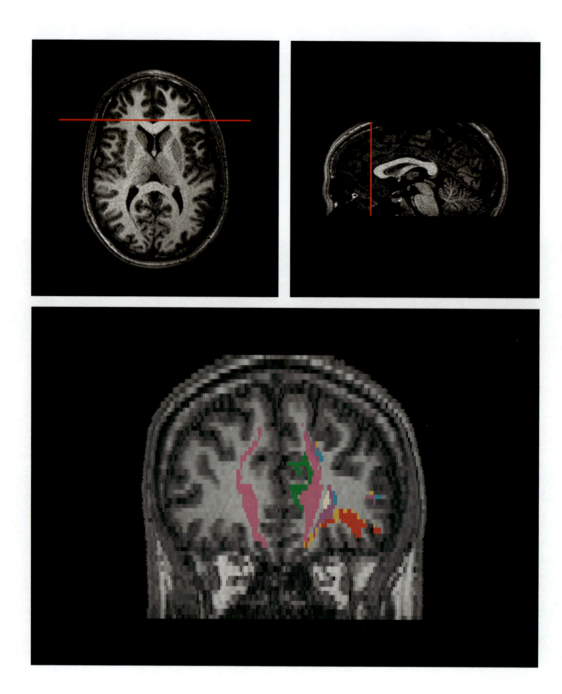

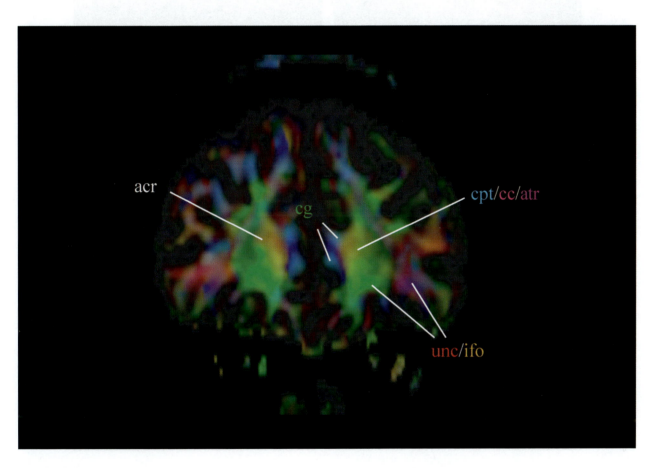

acr: anterior corona radiata
atr: anterior thalamic radiation
cc: corpus callosum
cg: cingulum

cpt: corticopontine tract
ifo: inferior fronto-occipital fasciculus
unc: uncinate fasciculus

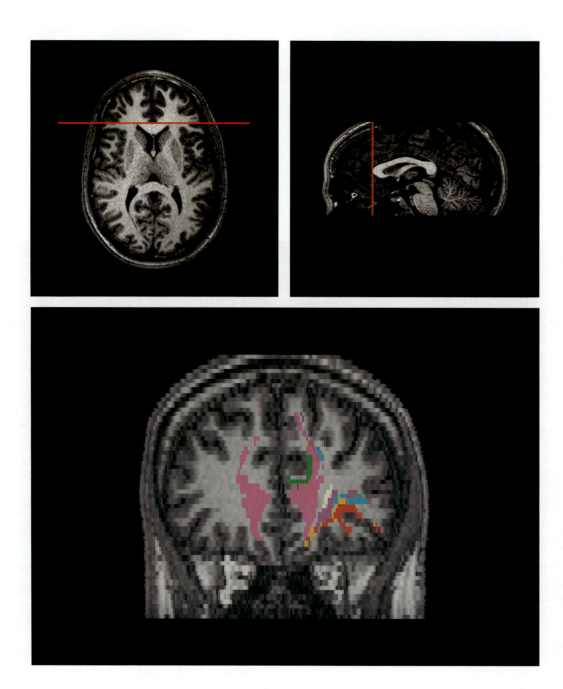

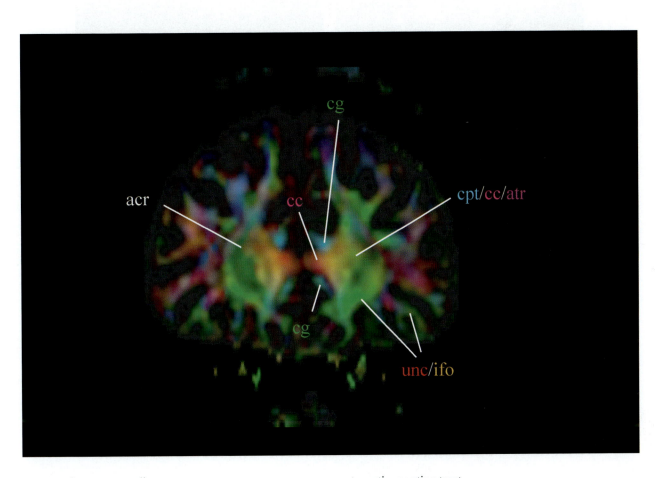

acr: anterior corona radiata
atr: anterior thalamic radiation
cc: corpus callosum
cg: cingulum

cpt: corticopontine tract
ifo: inferior fronto-occipital fasciculus
unc: uncinate fasciculus

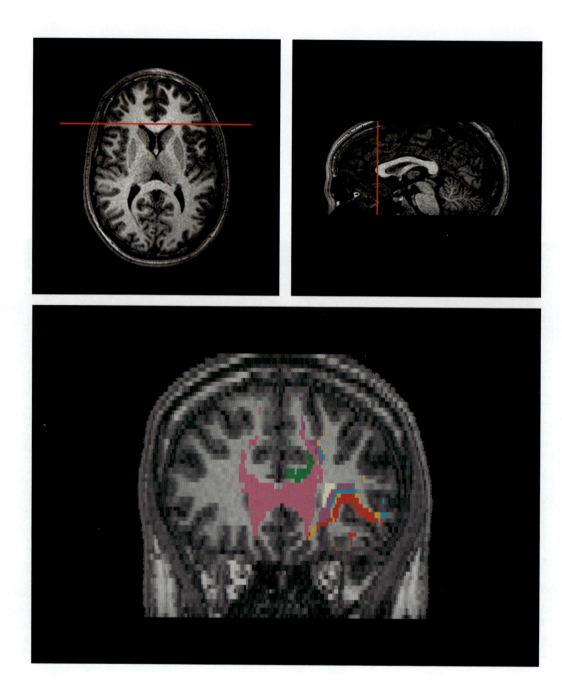

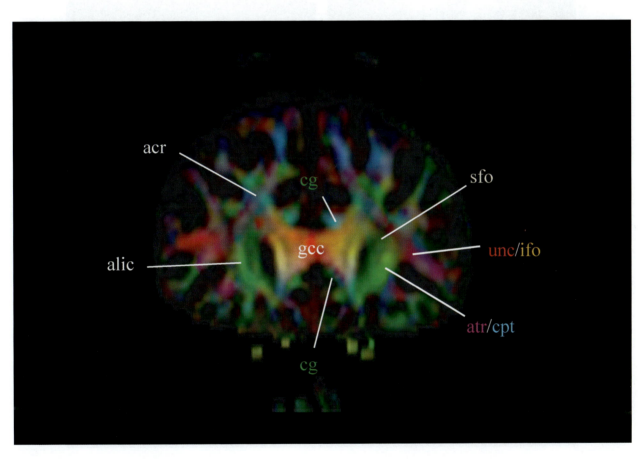

acr: anterior corona radiata
alic: anterior limb of the internal capsule
atr: anterior thalamic radiation
cg: cingulum
cpt: corticopontine tract

gcc: genu of the corpus callosum
ifo: inferior fronto-occipital fasciculus
sfo: superior fronto-occipital fasciculus
unc: uncinate fasciculus

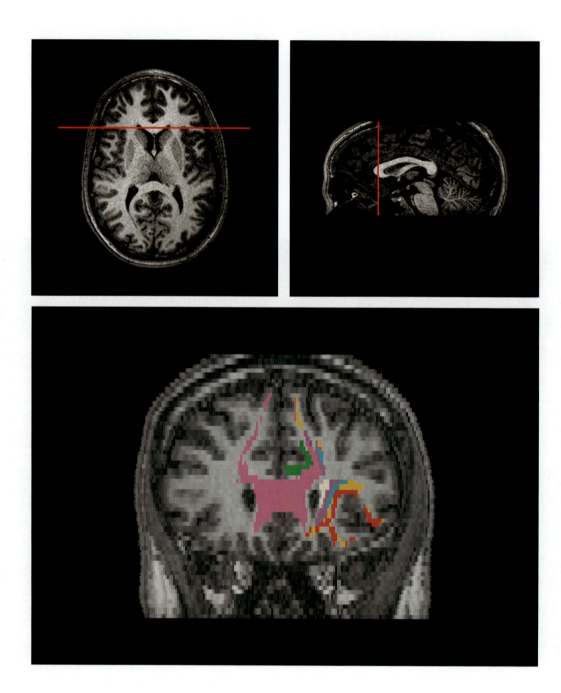

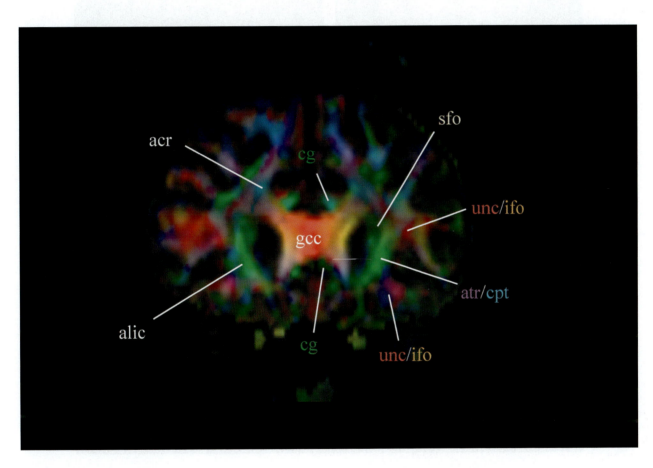

acr: anterior corona radiata
alic: anterior limb of the internal capsule
atr: anterior thalamic radiation
cg: cingulum
cpt: corticopontine tract

gcc: genu of the corpus callosum
ifo: inferior fronto-occipital fasciculus
sfo: superior fronto-occipital fasciculus
unc: uncinate fasciculus

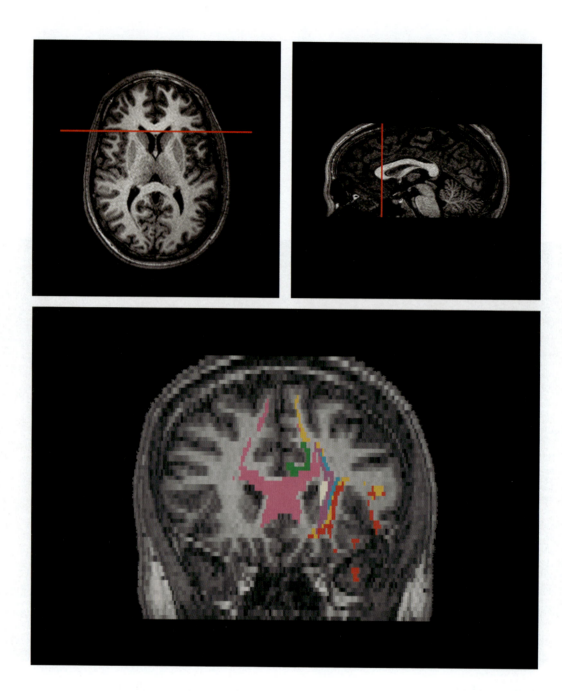

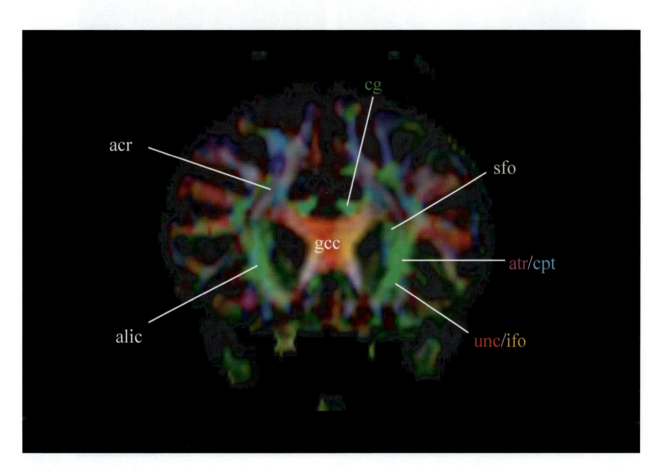

acr: anterior corona radiata
alic: anterior limb of the internal capsule
atr: anterior thalamic radiation
cg: cingulum
cpt: corticopontine tract

gcc: genu of the corpus callosum
ifo: inferior fronto-occipital fasciculus
sfo: superior fronto-occipital fasciculus
unc: uncinate fasciculus

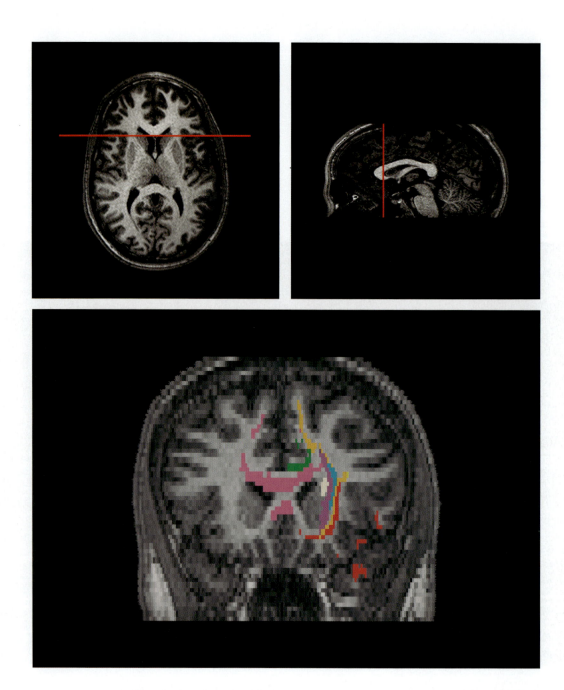

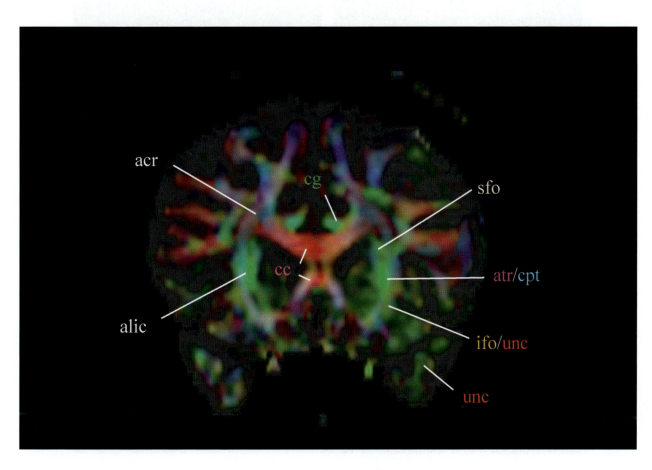

acr: anterior corona radiata
alic: anterior limb of the internal capsule
atr: anterior thalamic radiation
cc: corpus callosum
cg: cingulum

cpt: corticopontine tract
ifo: inferior fronto-occipital fasciculus
sfo: superior fronto-occipital fasciculus
unc: uncinate fasciculus

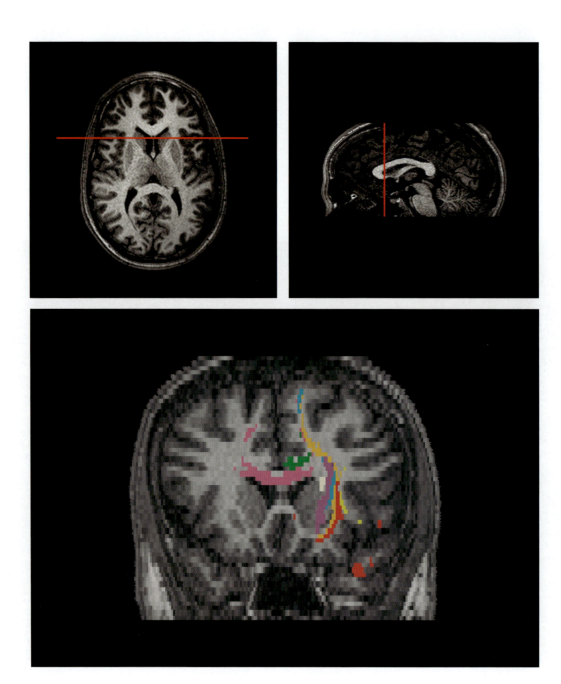

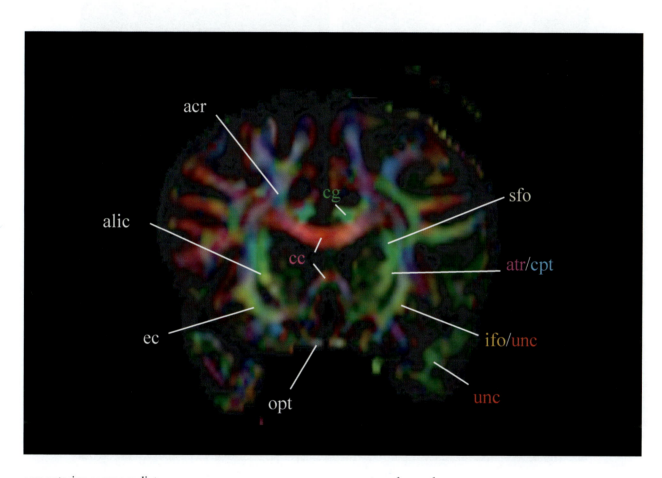

acr: anterior corona radiata
alic: anterior limb of the internal capsule
atr: anterior thalamic radiation
cc: corpus callosum
cg: cingulum
cpt: corticopontine tract

ec: external capsule
ifo: inferior fronto-occipital fasciculus
opt: optic tract
sfo: superior fronto-occipital fasciculus
unc: uncinate fasciculus

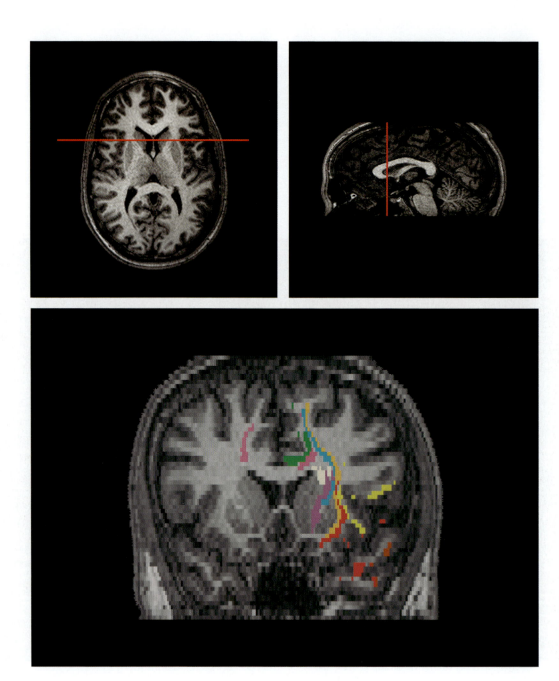

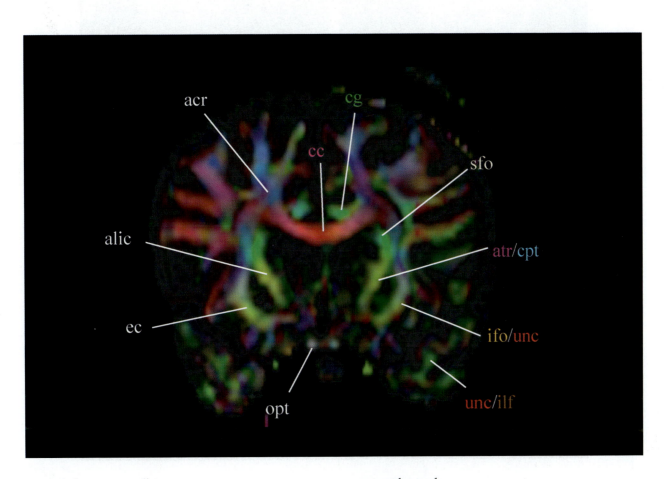

acr: anterior corona radiata
alic: anterior limb of the internal capsule
atr: anterior thalamic radiation
cc: corpus callosum
cg: cingulum
cpt: corticopontine tract

ec: external capsule
ifo: inferior fronto-occipital fasciculus
ilf: inferior longitudinal faciculus
opt: optic tract
sfo: superior fronto-occipital fasciculus
unc: uncinate fasciculus

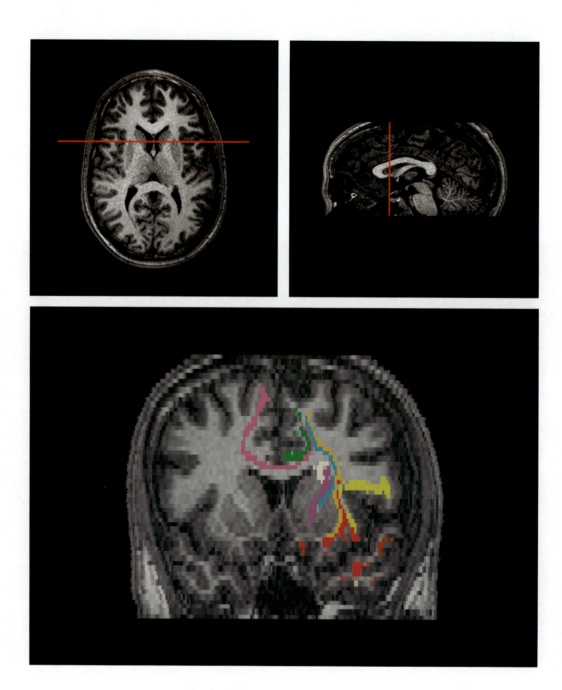

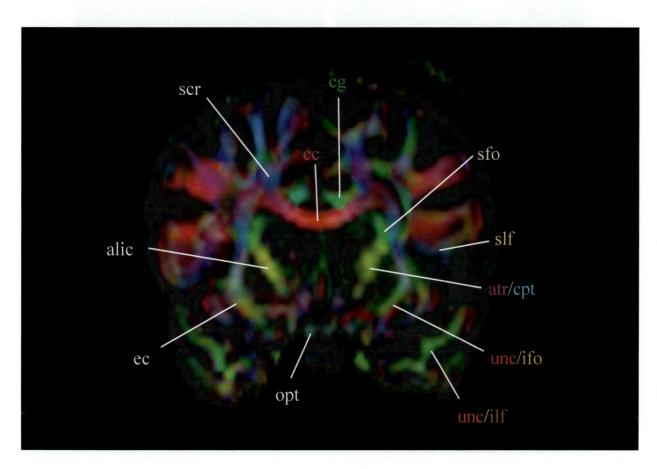

alic: anterior limb of the internal capsule
atr: anterior thalamic radiation
cc: corpus callosum
cg: cingulum
cpt: corticopontine tract
ec: external capsule
ifo: inferior fronto-occipital fasciculus

ilf: inferior longitudinal faciculus
opt: optic tract
scr: superior corona radiata
sfo: superior fronto-occipital fasciculus
slf: superior longitudinal fasciculus
unc: uncinate fasciculus

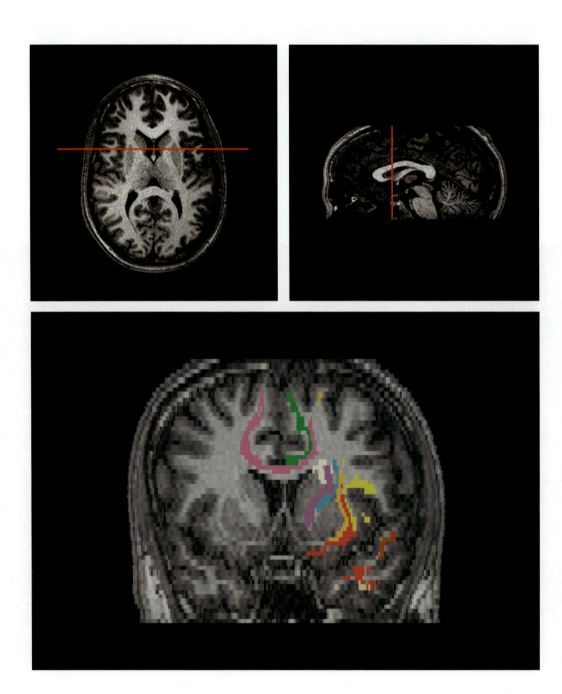

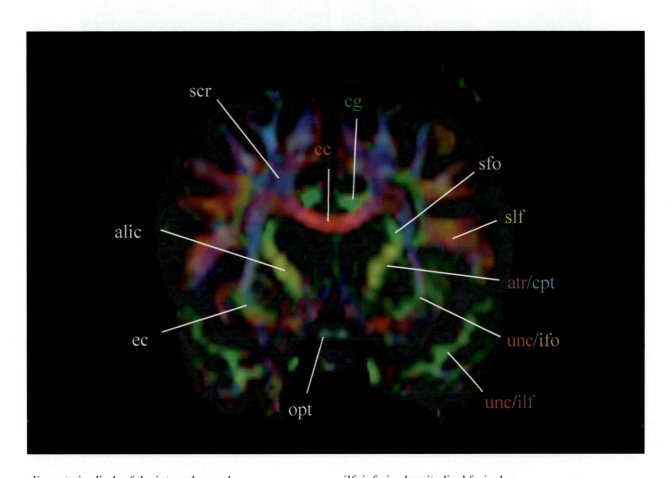

alic: anterior limb of the internal capsule
atr: anterior thalamic radiation
cc: corpus callosum
cg: cingulum
cpt: corticopontine tract
ec: external capsule
ifo: inferior fronto-occipital fasciculus

ilf: inferior longitudinal faciculus
opt: optic tract
scr: superior corona radiata
sfo: superior fronto-occipital fasciculus
slf: superior longitudinal fasciculus
unc: uncinate fasciculus

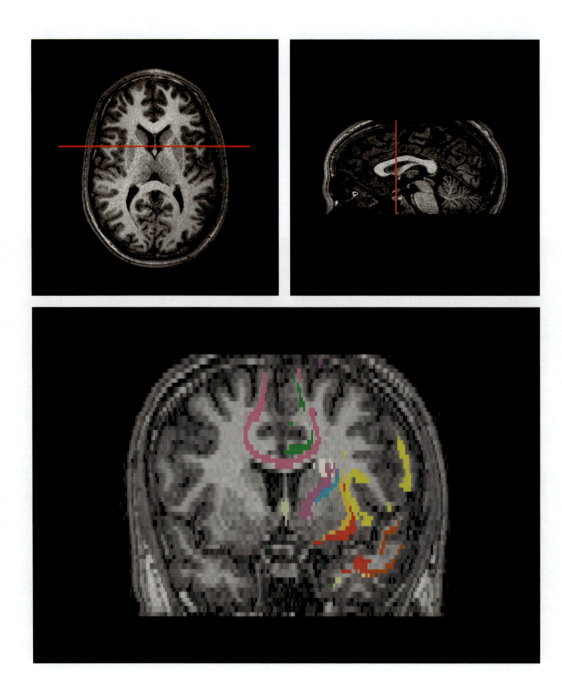

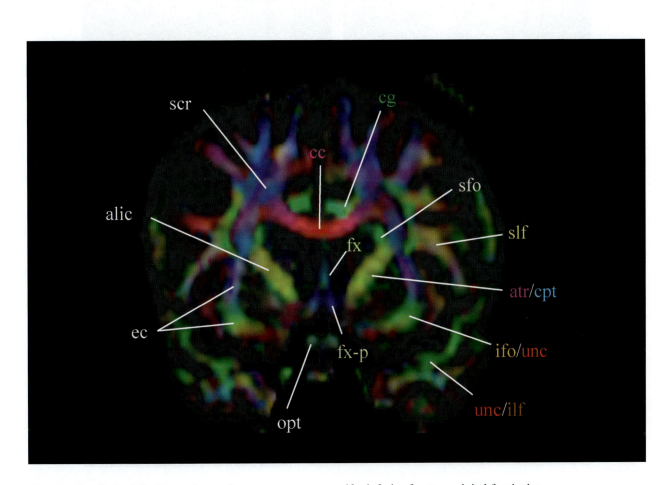

alic: anterior limb of the internal capsule
atr: anterior thalamic radiation
cc: corpus callosum
cg: cingulum
cpt: corticopontine tract
ec: external capsule
fx: fornix
fx-p: precommissural part of the fornix

ifo: inferior fronto-occipital fasciculus
ilf: inferior longitudinal faciculus
opt: optic tract
scr: superior corona radiata
sfo: superior fronto-occipital fasciculus
slf: superior longitudinal fasciculus
unc: uncinate fasciculus

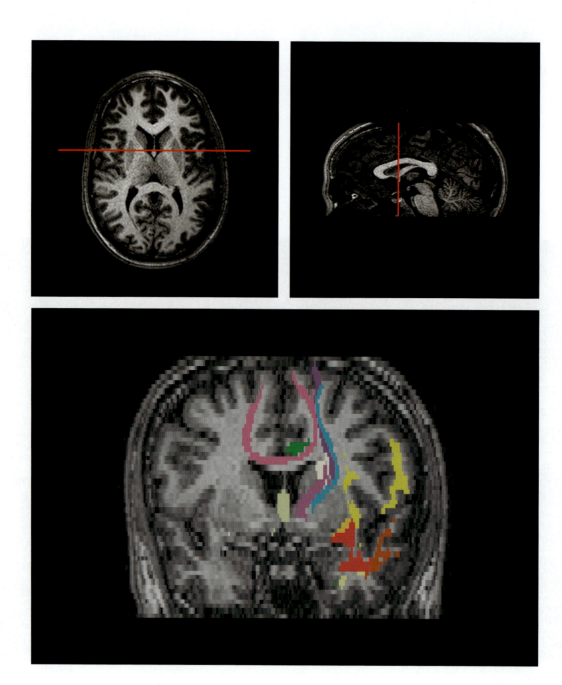

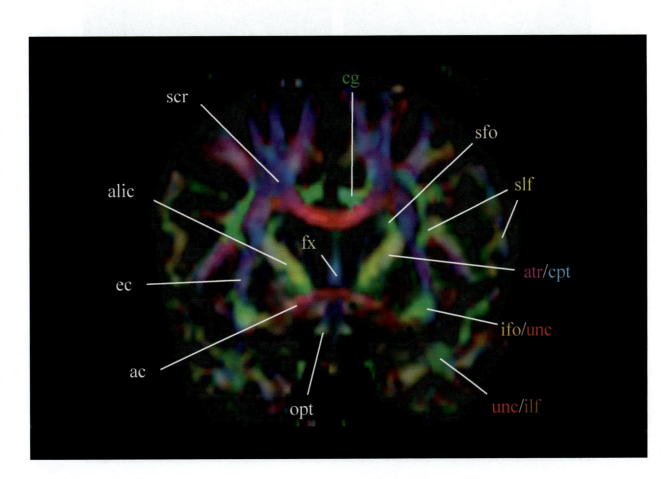

ac: anterior commissure
alic: anterior limb of the internal capsule
atr: anterior thalamic radiation
cg: cingulum
cpt: corticopontine tract
ec: external capsule
fx: fornix

ifo: inferior fronto-occipital fasciculus
ilf: inferior longitudinal faciculus
opt: optic tract
scr: superior corona radiata
sfo: superior fronto-occipital fasciculus
slf: superior longitudinal fasciculus
unc: uncinate fasciculus

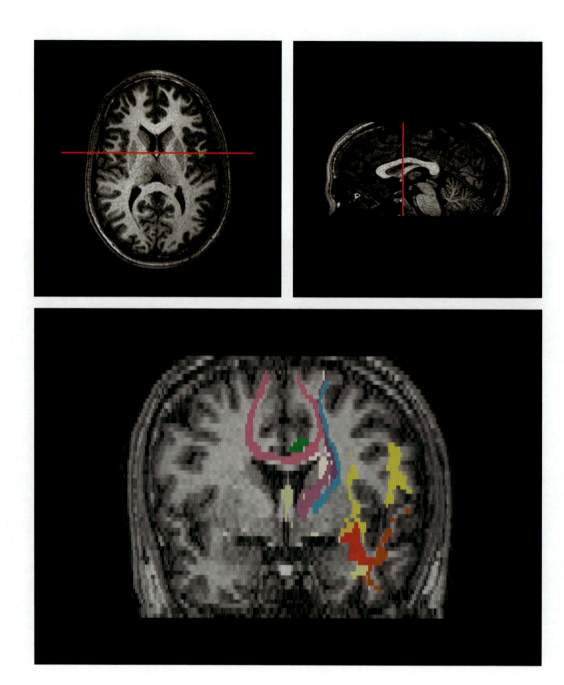

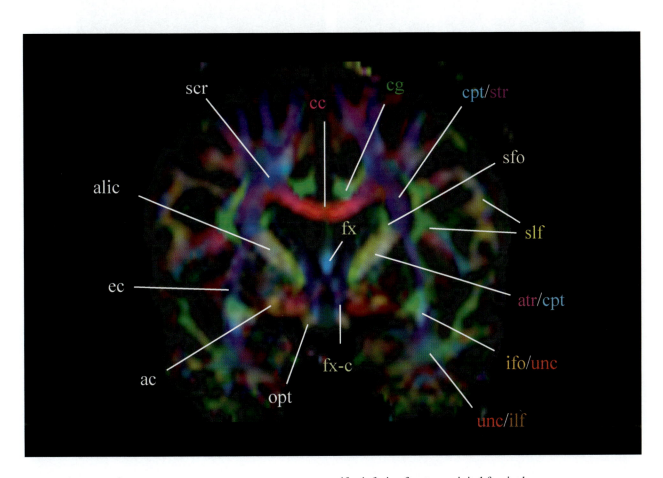

ac: anterior commissure
alic: anterior limb of the internal capsule
atr: anterior thalamic radiation
cc: corpus callosum
cg: cingulum
cpt: corticopontine tract
ec: external capsule
fx: fornix
fx-c: column of the fornix

ifo: inferior fronto-occipital fasciculus
ilf: inferior longitudinal faciculus
opt: optic tract
scr: superior corona radiata
sfo: superior fronto-occipital fasciculus
slf: superior longitudinal fasciculus
str: superior thalamic radiation
unc: uncinate fasciculus

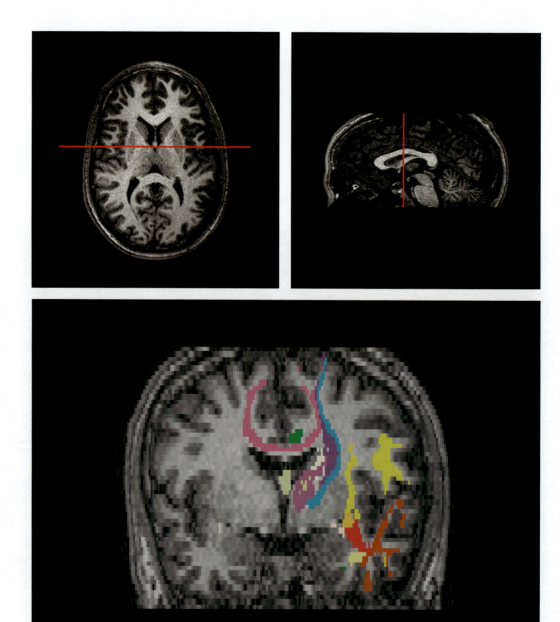

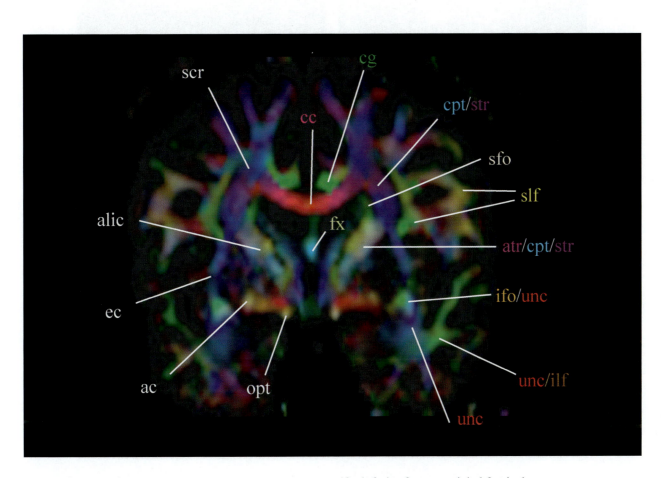

ac: anterior commissure
alic: anterior limb of the internal capsule
atr: anterior thalamic radiation
cc: corpus callosum
cg: cingulum
cpt: corticopontine tract
ec: external capsule
fx: fornix

ifo: inferior fronto-occipital fasciculus
ilf: inferior longitudinal faciculus
opt: optic tract
scr: superior corona radiata
sfo: superior fronto-occipital fasciculus
slf: superior longitudinal fasciculus
str: superior thalamic radiation
unc: uncinate fasciculus

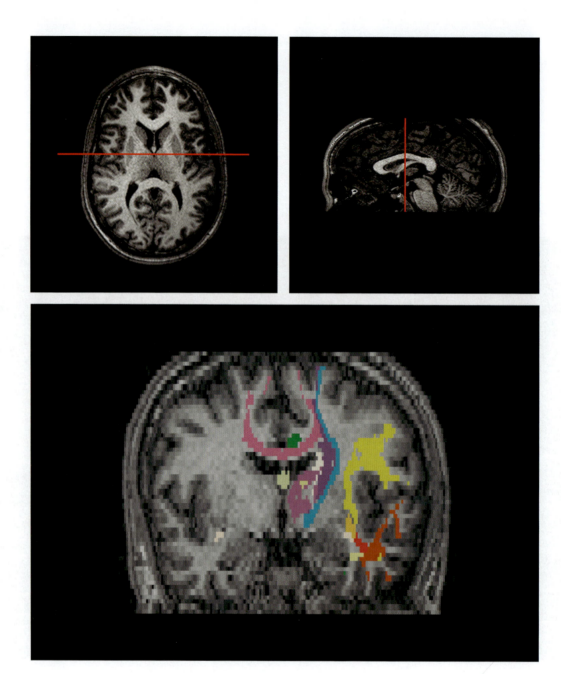

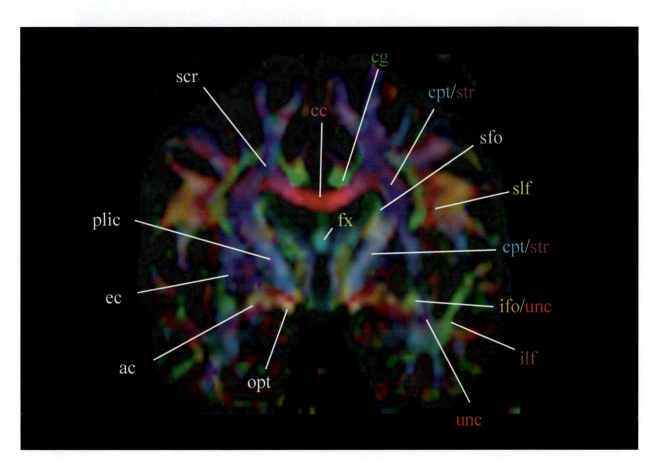

ac: anterior commissure
cc: corpus callosum
cg: cingulum
cpt: corticopontine tract
ec: external capsule
fx: fornix
ifo: inferior fronto-occipital fasciculus
ilf: inferior longitudinal faciculus

opt: optic tract
plic: posterior limb of the internal capsule
scr: superior corona radiata
sfo: superior fronto-occipital fasciculus
slf: superior longitudinal fasciculus
str: superior thalamic radiation
unc: uncinate fasciculus

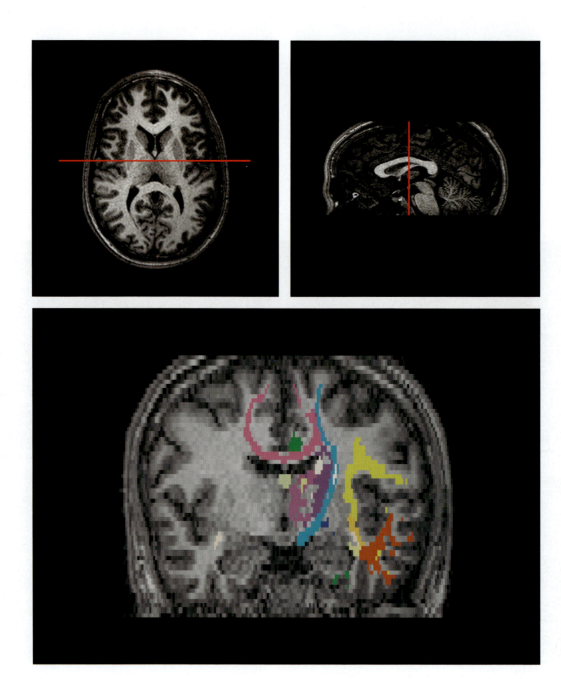

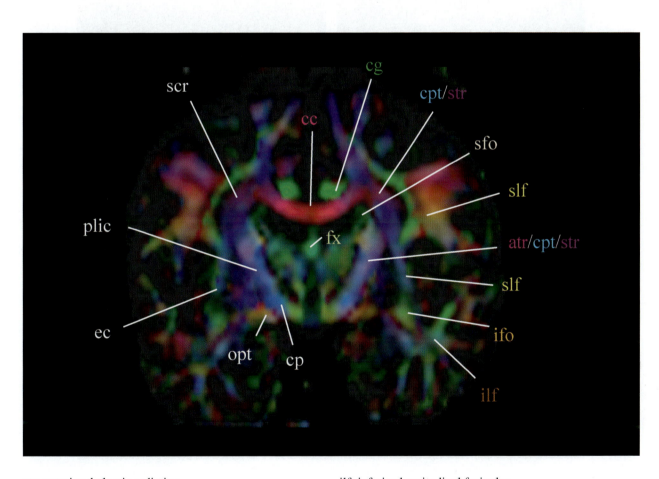

atr: anterior thalamic radiation
cc: corpus callosum
cg: cingulum
cp: cerebral peduncle
cpt: corticopontine tract
ec: external capsule
fx: fornix
ifo: inferior fronto-occipital fasciculus

ilf: inferior longitudinal faciculus
opt: optic tract
plic: posterior limb of the internal capsule
scr: superior corona radiata
sfo: superior fronto-occipital fasciculus
slf: superior longitudinal fasciculus
str: superior thalamic radiation

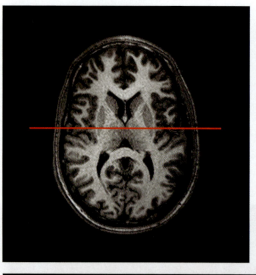
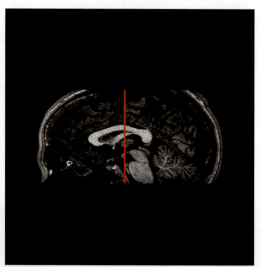
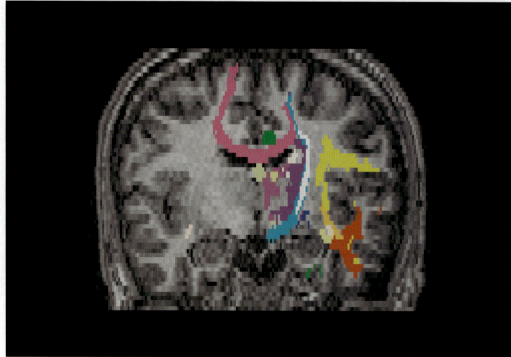

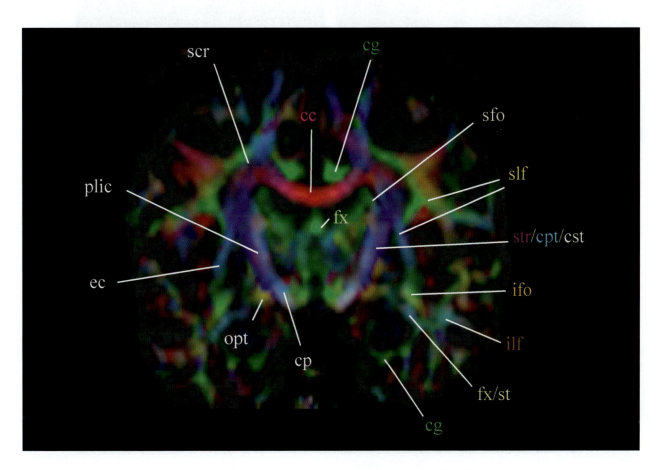

cc: corpus callosum
cg: cingulum
cp: cerebral peduncle
cpt: corticopontine tract
cst: corticospinal tract
ec: external capsule
fx: fornix
ifo: inferior fronto-occipital fasciculus

ilf: inferior longitudinal faciculus
opt: optic tract
plic: posterior limb of the internal capsule
scr: superior corona radiata
sfo: superior fronto-occipital fasciculus
slf: superior longitudinal fasciculus
st: stria terminalis
str: superior thalamic radiation

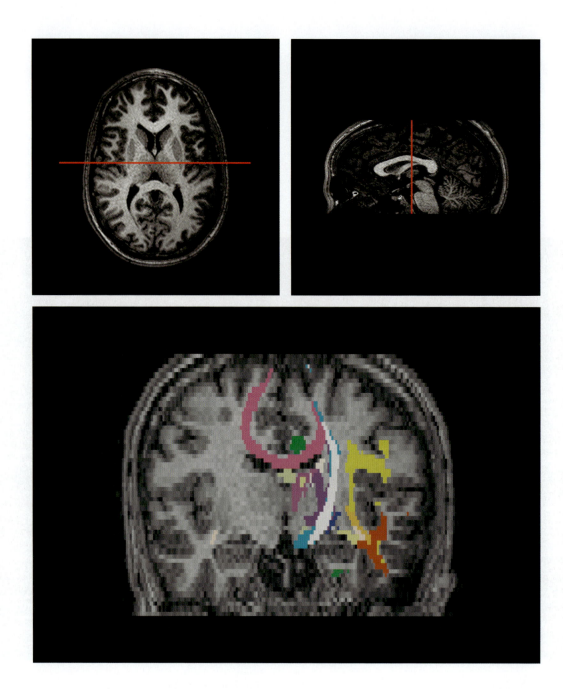

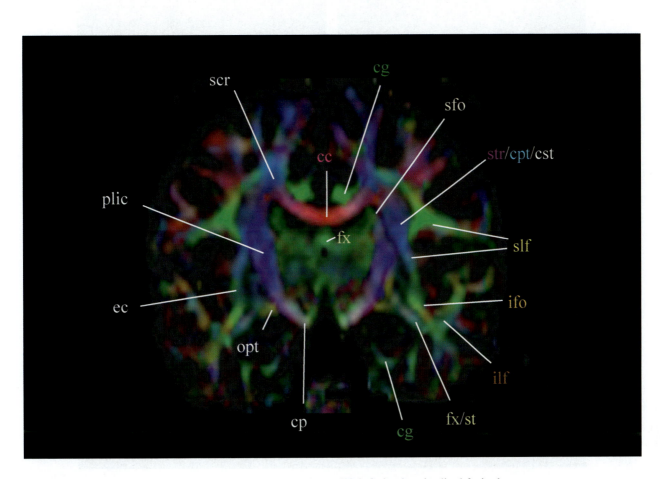

cc: corpus callosum
cg: cingulum
cp: cerebral peduncle
cpt: corticopontine tract
cst: corticospinal tract
ec: external capsule
fx: fornix
ifo: inferior fronto-occipital fasciculus

ilf: inferior longitudinal faciculus
opt: optic tract
plic: posterior limb of the internal capsule
scr: superior corona radiata
sfo: superior fronto-occipital fasciculus
slf: superior longitudinal fasciculus
st: stria terminalis
str: superior thalamic radiation

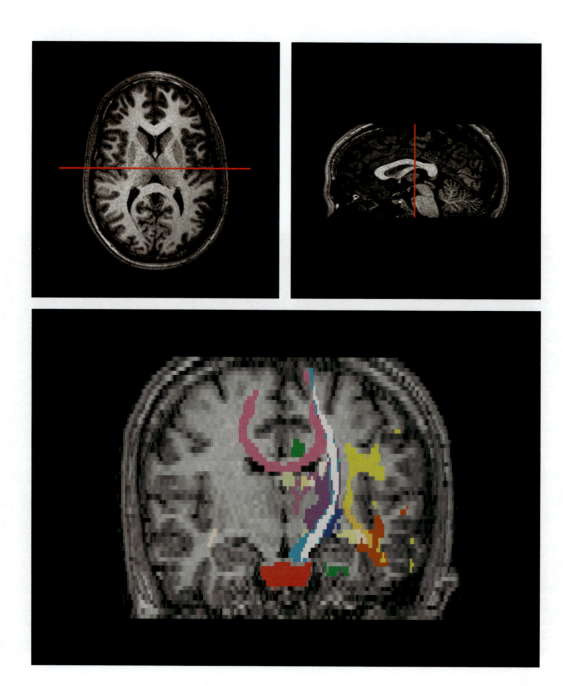

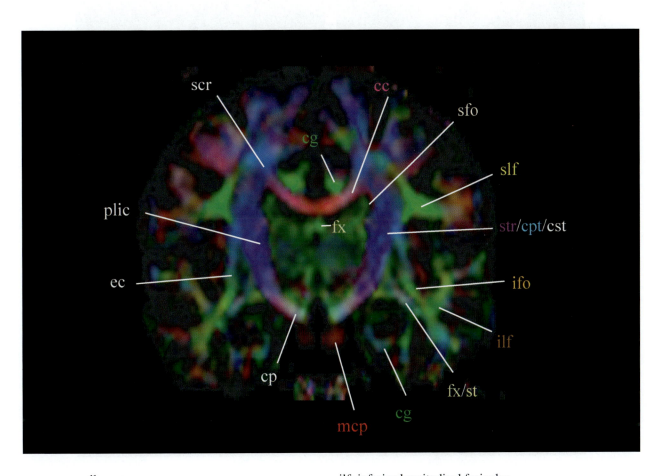

cc: corpus callosum
cg: cingulum
cp: cerebral peduncle
cpt: corticopontine tract
cst: corticospinal tract
ec: external capsule
fx: fornix
ifo: inferior fronto-occipital fasciculus

ilf: inferior longitudinal faciculus
mcp: middle cerebellar peduncle
plic: posterior limb of the internal capsule
scr: superior corona radiata
sfo: superior fronto-occipital fasciculus
slf: superior longitudinal fasciculus
st: stria terminalis
str: superior thalamic radiation

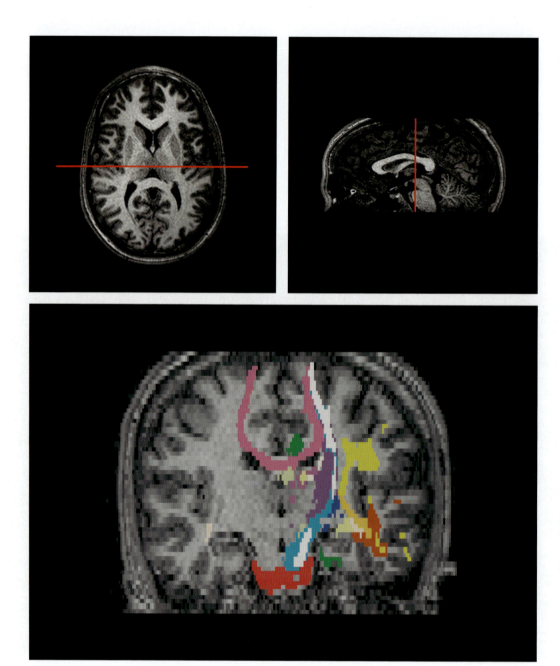

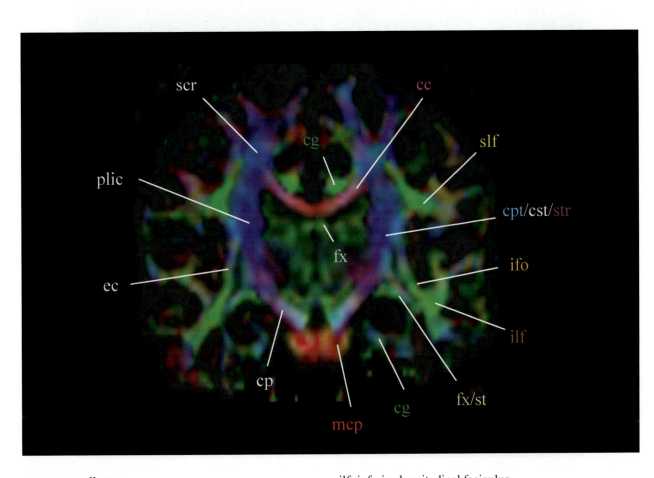

cc: corpus callosum
cg: cingulum
cp: cerebral peduncle
cpt: corticopontine tract
cst: corticospinal tract
ec: external capsule
fx: fornix
ifo: inferior fronto-occipital fasciculus

ilf: inferior longitudinal faciculus
mcp: middle cerebellar peduncle
plic: posterior limb of the internal capsule
scr: superior corona radiata
slf: superior longitudinal fasciculus
st: stria terminalis
str: superior thalamic radiation

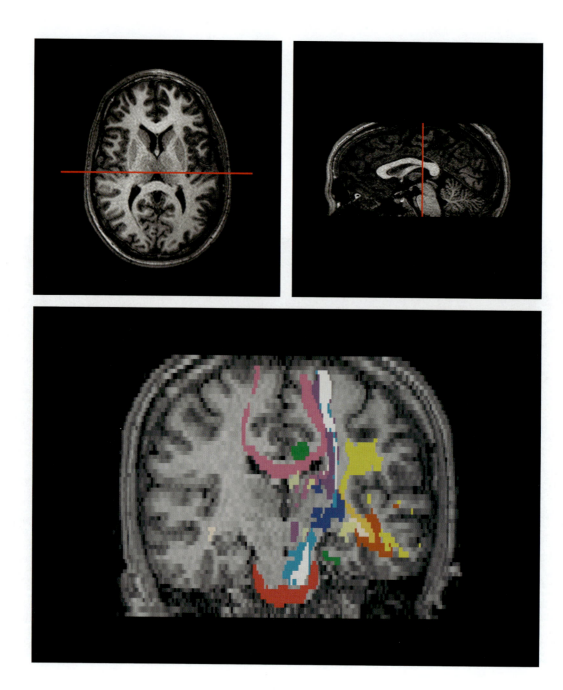

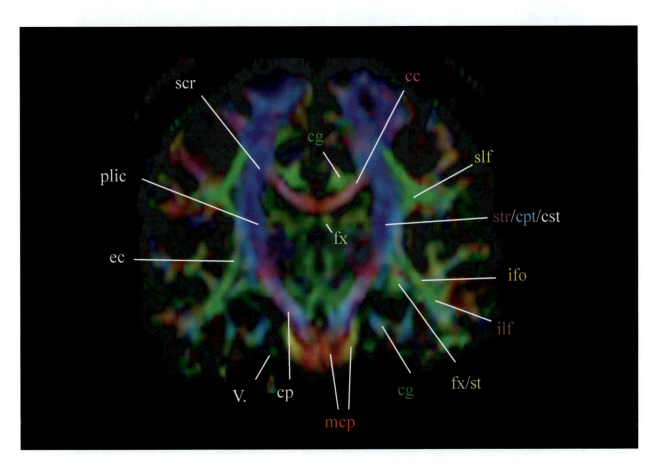

cc: corpus callosum
cg: cingulum
cp: cerebral peduncle
cpt: corticopontine tract
cst: corticospinal tract
ec: external capsule
fx: fornix
ifo: inferior fronto-occipital fasciculus

ilf: inferior longitudinal faciculus
mcp: middle cerebellar peduncle
plic: posterior limb of the internal capsule
scr: superior corona radiata
slf: superior longitudinal fasciculus
st: stria terminalis
str: superior thalamic radiation
V.: fifth cranial nerve (trigeminal nerve)

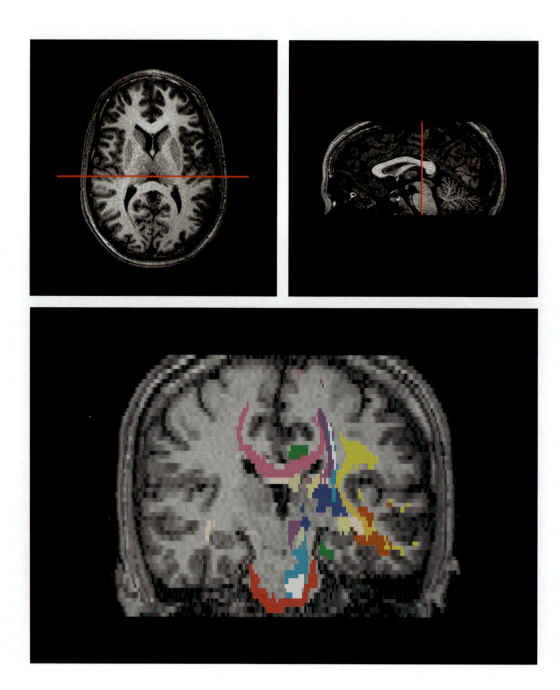

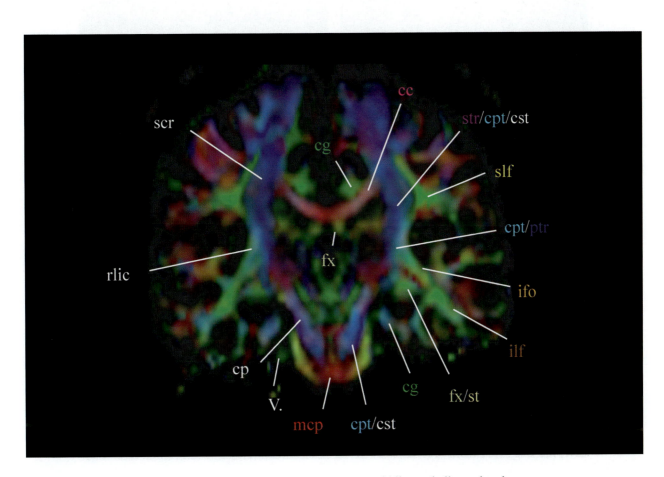

cc: corpus callosum
cg: cingulum
cp: cerebral peduncle
cpt: corticopontine tract
cst: corticospinal tract
fx: fornix
ifo: inferior fronto-occipital fasciculus
ilf: inferior longitudinal faciculus

mcp: middle cerebellar peduncle
ptr: posterior thalamic radiation
rlic: retrolenticular part of the internal capsule
scr: superior corona radiata
slf: superior longitudinal fasciculus
st: stria terminalis
str: superior thalamic radiation
V.: fifth cranial nerve (trigeminal nerve)

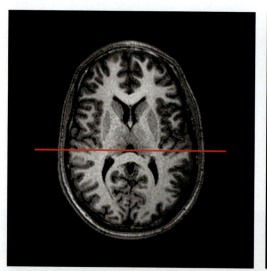
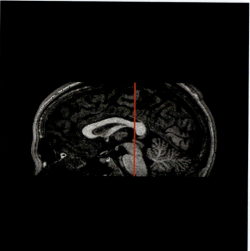
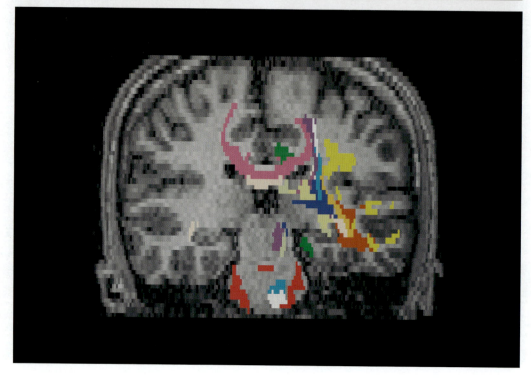

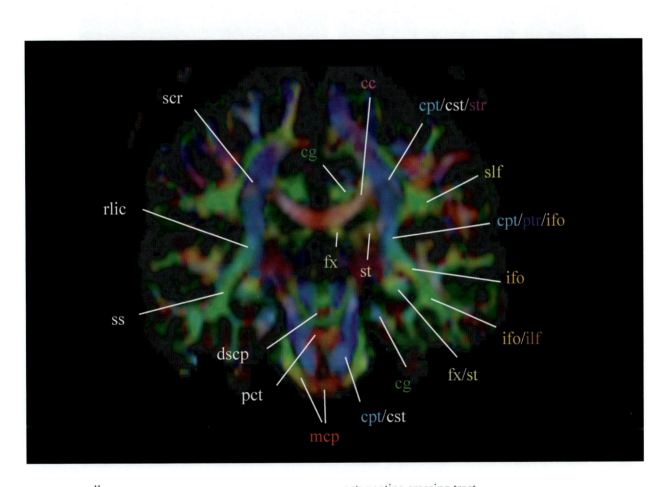

cc: corpus callosum
cg: cingulum
cpt: corticopontine tract
cst: corticospinal tract
dscp: decussation of the superior cerebellar peduncles
fx: fornix
ifo: inferior fronto-occipital fasciculus
ilf: inferior longitudinal faciculus
mcp: middle cerebellar peduncle

pct: pontine crossing tract
ptr: posterior thalamic radiation
rlic: retrolenticular part of the internal capsule
scr: superior corona radiata
slf: superior longitudinal fasciculus
ss: sagittal stratum
st: stria terminalis
str: superior thalamic radiation

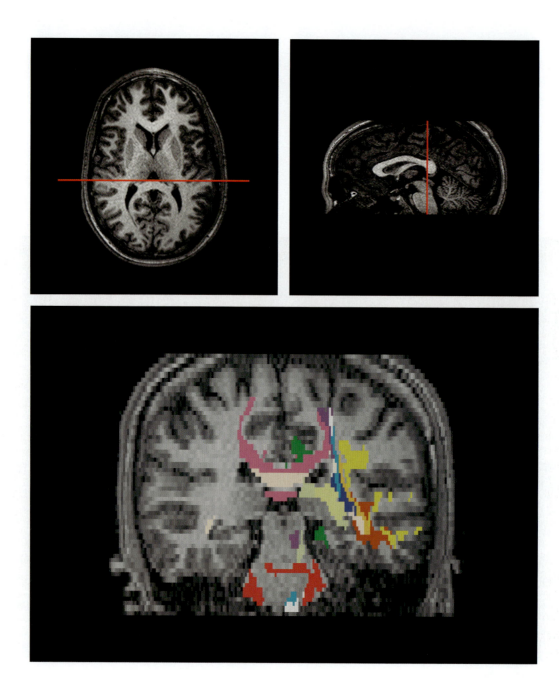

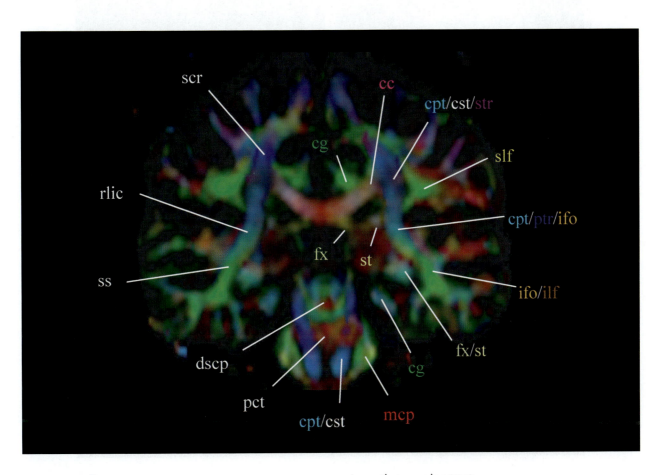

cc: corpus callosum
cg: cingulum
cpt: corticopontine tract
cst: corticospinal tract
dscp: decussation of the superior cerebellar peduncles
fx: fornix
ifo: inferior fronto-occipital fasciculus
ilf: inferior longitudinal faciculus
mcp: middle cerebellar peduncle

pct: pontine crossing tract
ptr: posterior thalamic radiation
rlic: retrolenticular part of the internal capsule
scr: superior corona radiata
slf: superior longitudinal fasciculus
ss: sagittal stratum
st: stria terminalis
str: superior thalamic radiation

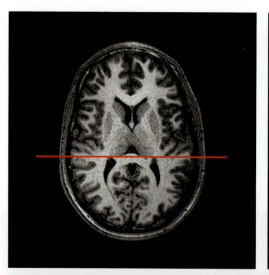
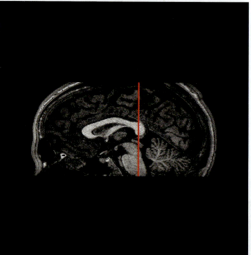
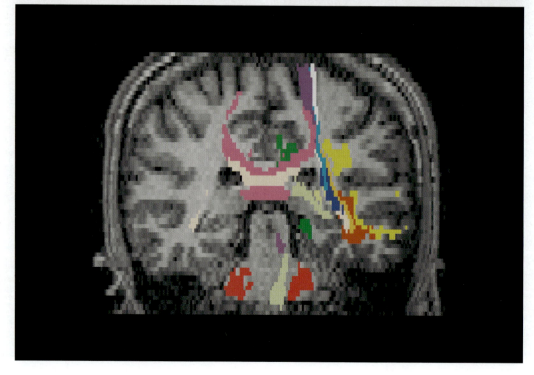

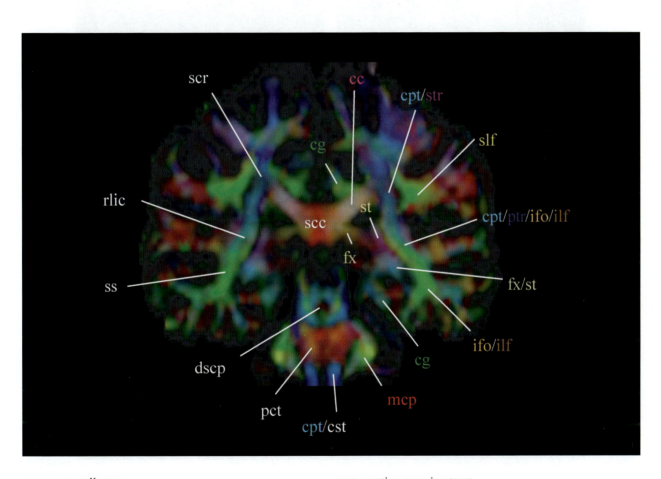

cc: corpus callosum
cg: cingulum
cpt: corticopontine tract
cst: corticospinal tract
dscp: decussation of the superior cerebellar peduncles
fx: fornix
ifo: inferior fronto-occipital fasciculus
ilf: inferior longitudinal faciculus
mcp: middle cerebellar peduncle

pct: pontine crossing tract
ptr: posterior thalamic radiation
rlic: retrolenticular part of the internal capsule
scc: splenium of the corpus callosum
scr: superior corona radiata
slf: superior longitudinal fasciculus
ss: sagittal stratum
st: stria terminalis
str: superior thalamic radiation

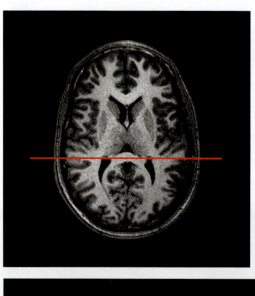
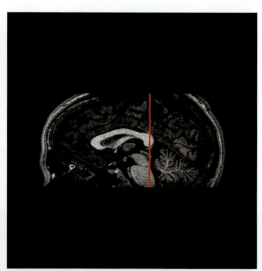
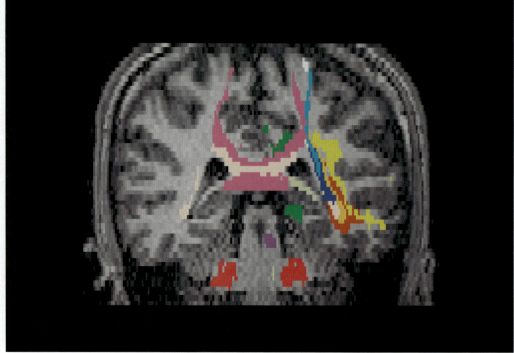

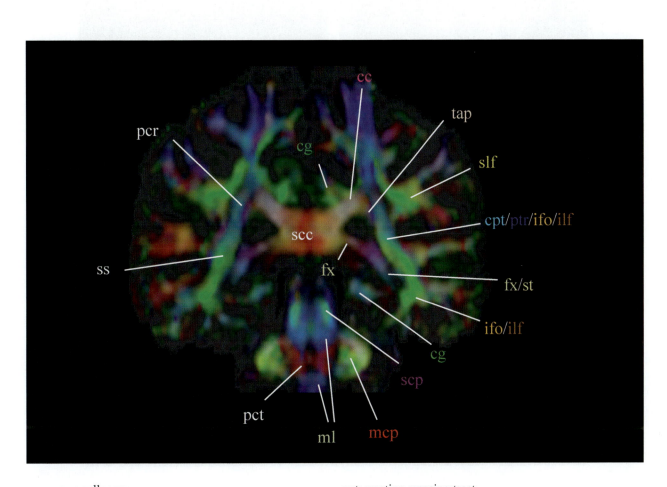

cc: corpus callosum
cg: cingulum
cpt: corticopontine tract
fx: fornix
ifo: inferior fronto-occipital fasciculus
ilf: inferior longitudinal faciculus
mcp: middle cerebellar peduncle
ml: medial lemniscus
pcr: posterior corona radiata

pct: pontine crossing tract
ptr: posterior thalamic radiation
scc: splenium of the corpus callosum
scp: superior cerebellar peduncle
slf: superior longitudinal fasciculus
ss: sagittal stratum
st: stria terminalis
tap: tapetum

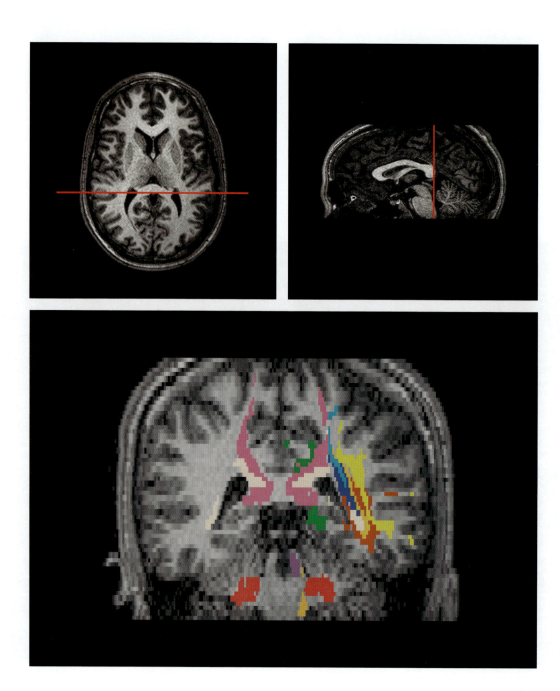

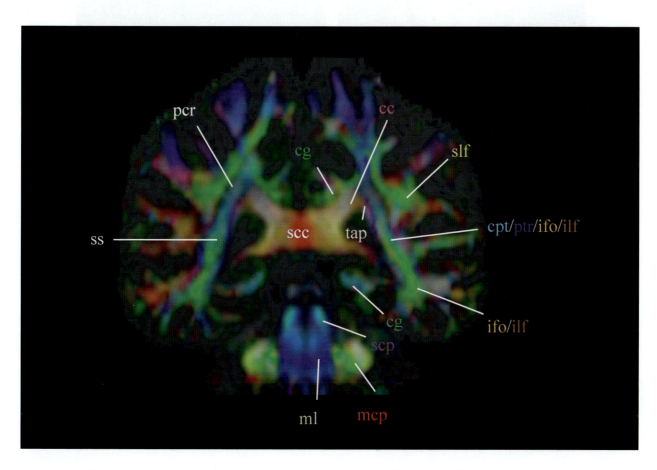

cc: corpus callosum
cg: cingulum
cpt: corticopontine tract
ifo: inferior fronto-occipital fasciculus
ilf: inferior longitudinal faciculus
mcp: middle cerebellar peduncle
ml: medial lemniscus

pcr: posterior corona radiata
ptr: posterior thalamic radiation
scc: splenium of the corpus callosum
scp: superior cerebellar peduncle
slf: superior longitudinal fasciculus
ss: sagittal stratum
tap: tapetum

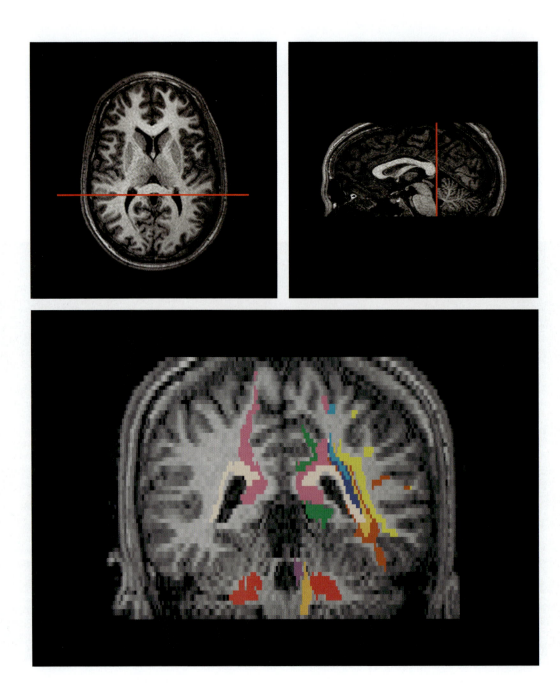

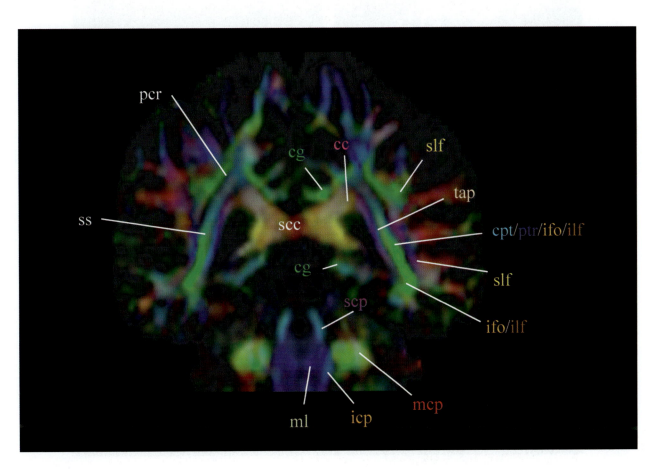

cc: corpus callosum
cg: cingulum
cpt: corticopontine tract
icp: inferior cerebellar peduncle
ifo: inferior fronto-occipital fasciculus
ilf: inferior longitudinal faciculus
mcp: middle cerebellar peduncle
ml: medial lemniscus

pcr: posterior corona radiata
ptr: posterior thalamic radiation
scc: splenium of the corpus callosum
scp: superior cerebellar peduncle
slf: superior longitudinal fasciculus
ss: sagittal stratum
tap: tapetum

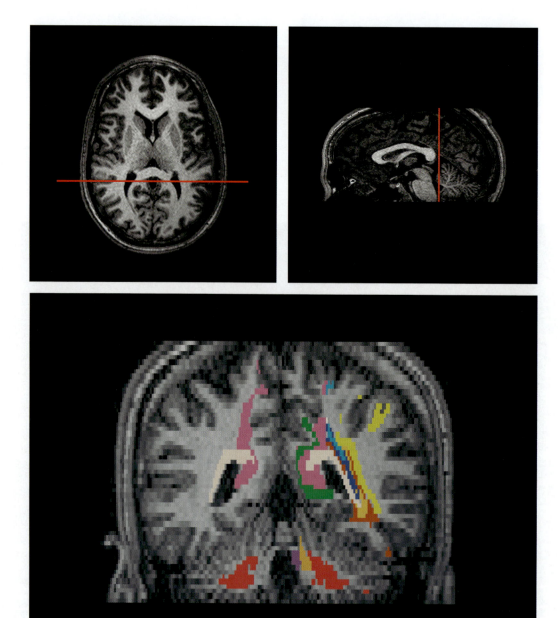

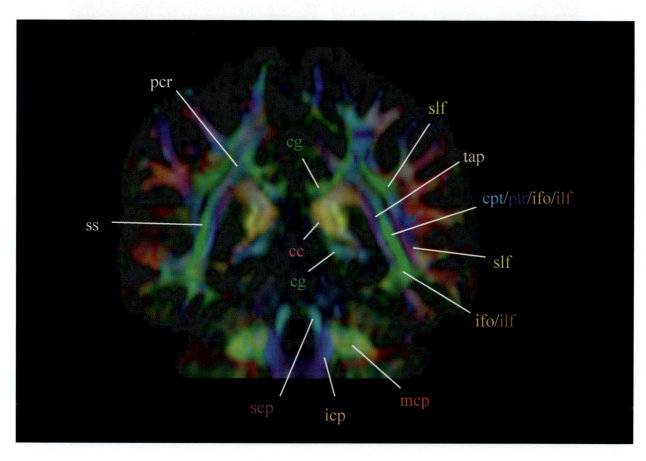

cc: corpus callosum
cg: cingulum
cpt: corticopontine tract
icp: inferior cerebellar peduncle
ifo: inferior fronto-occipital fasciculus
ilf: inferior longitudinal faciculus
mcp: middle cerebellar peduncle

pcr: posterior corona radiata
ptr: posterior thalamic radiation
scp: superior cerebellar peduncle
slf: superior longitudinal fasciculus
ss: sagittal stratum
tap: tapetum

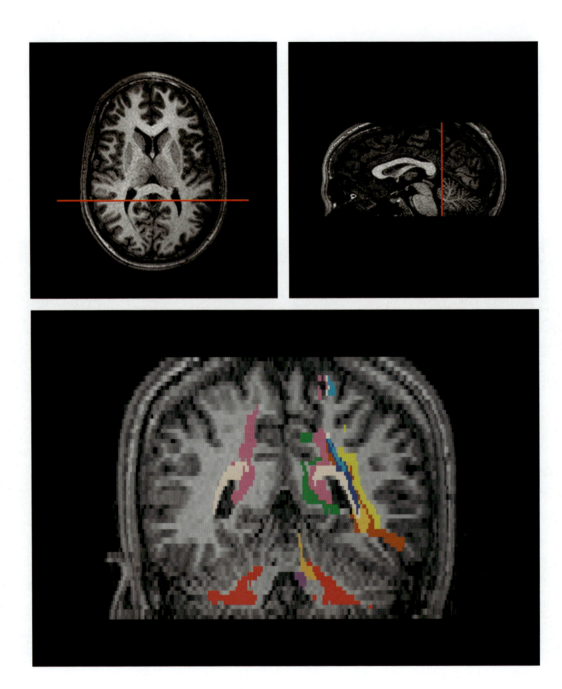

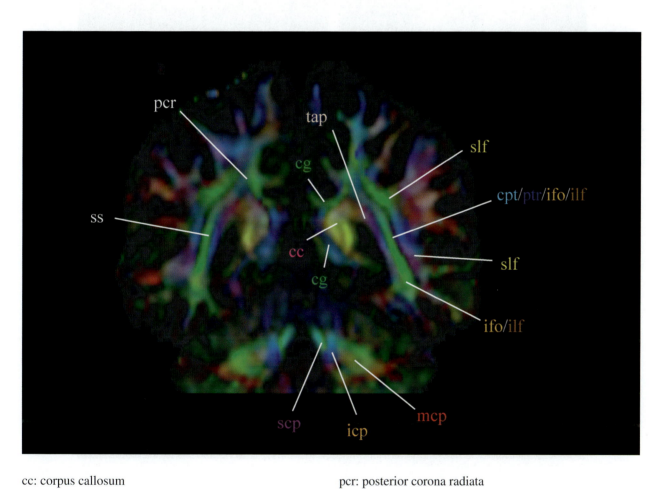

cc: corpus callosum
cg: cingulum
cpt: corticopontine tract
icp: inferior cerebellar peduncle
ifo: inferior fronto-occipital fasciculus
ilf: inferior longitudinal faciculus
mcp: middle cerebellar peduncle

pcr: posterior corona radiata
ptr: posterior thalamic radiation
scp: superior cerebellar peduncle
slf: superior longitudinal fasciculus
ss: sagittal stratum
tap: tapetum

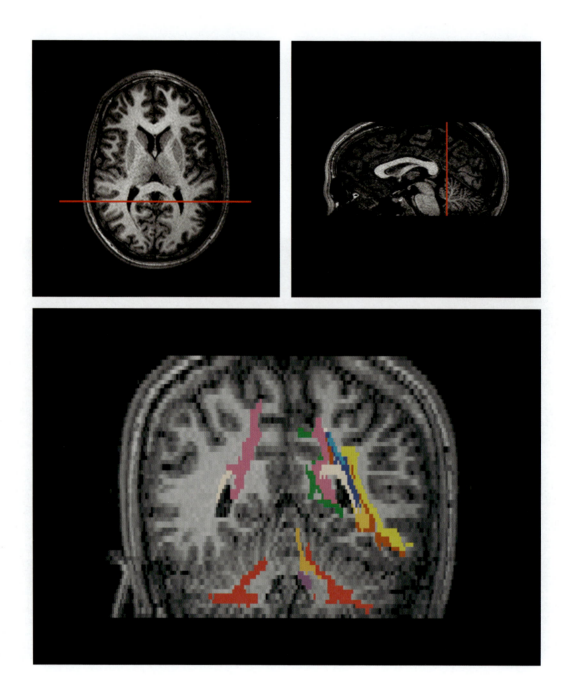

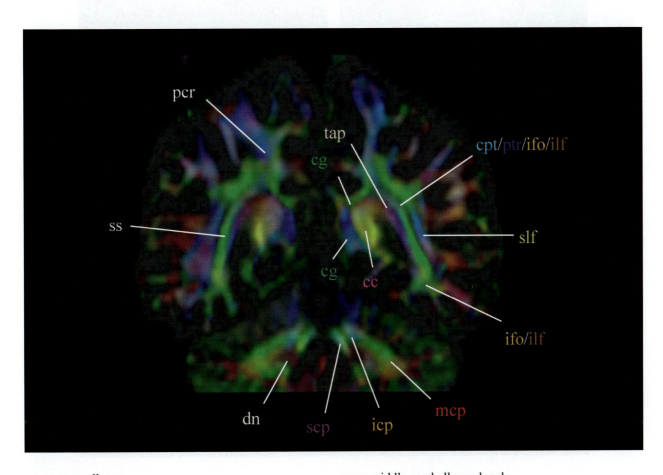

cc: corpus callosum
cg: cingulum
cpt: corticopontine tract
dn: dentate nucleus
icp: inferior cerebellar peduncle
ifo: inferior fronto-occipital fasciculus
ilf: inferior longitudinal faciculus

mcp: middle cerebellar peduncle
pcr: posterior corona radiata
ptr: posterior thalamic radiation
scp: superior cerebellar peduncle
slf: superior longitudinal fasciculus
ss: sagittal stratum
tap: tapetum

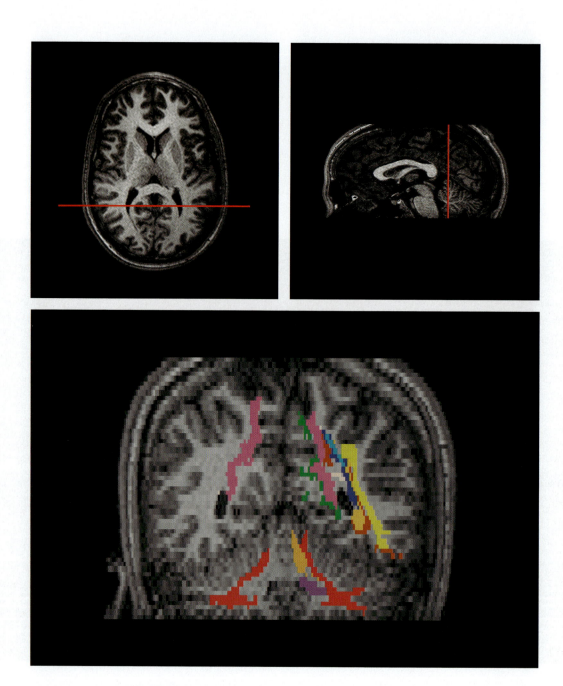

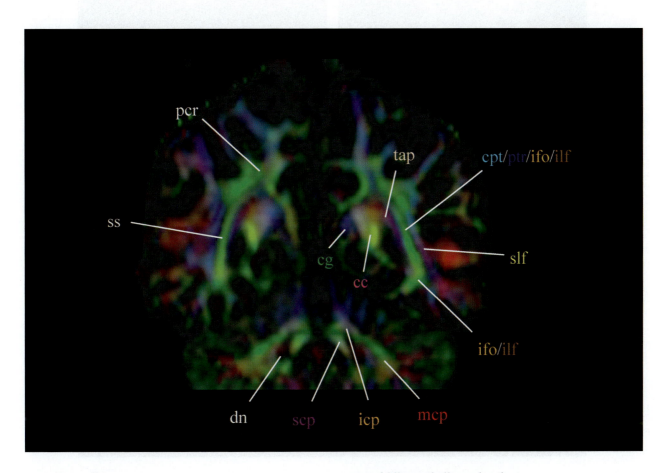

cc: corpus callosum
cg: cingulum
cpt: corticopontine tract
dn: dentate nucleus
icp: inferior cerebellar peduncle
ifo: inferior fronto-occipital fasciculus
ilf: inferior longitudinal faciculus

mcp: middle cerebellar peduncle
pcr: posterior corona radiata
ptr: posterior thalamic radiation
scp: superior cerebellar peduncle
slf: superior longitudinal fasciculus
ss: sagittal stratum
tap: tapetum

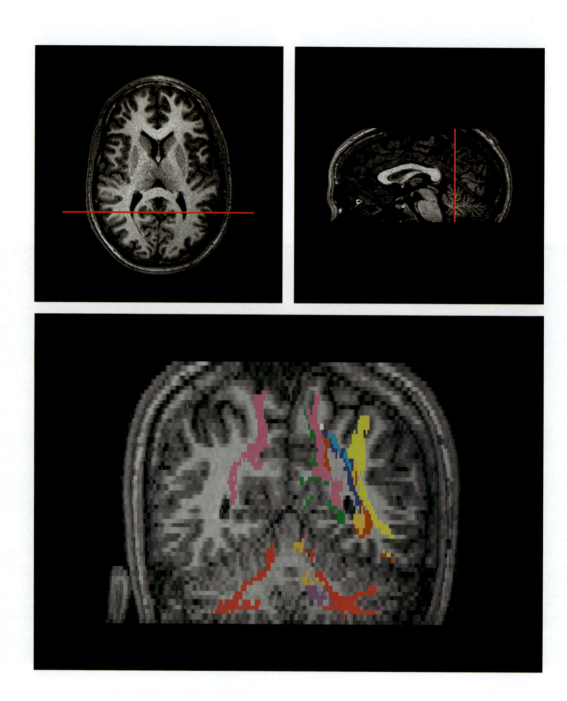

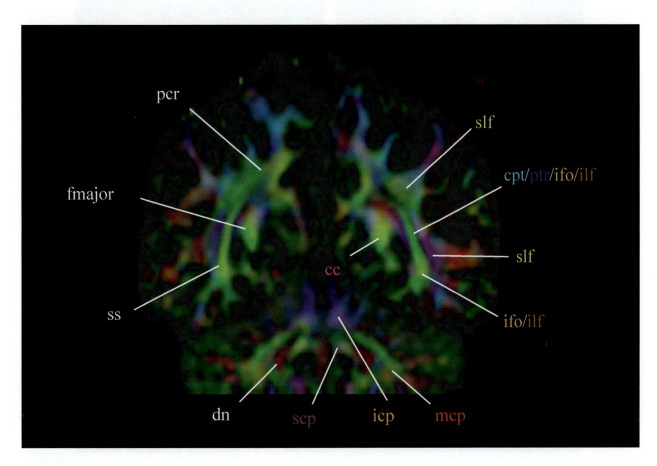

cc: corpus callosum
cpt: corticopontine tract
dn: dentate nucleus
fmajor: forceps major
icp: inferior cerebellar peduncle
ifo: inferior fronto-occipital fasciculus
ilf: inferior longitudinal faciculus

mcp: middle cerebellar peduncle
pcr: posterior corona radiata
ptr: posterior thalamic radiation
scp: superior cerebellar peduncle
slf: superior longitudinal fasciculus
ss: sagittal stratum

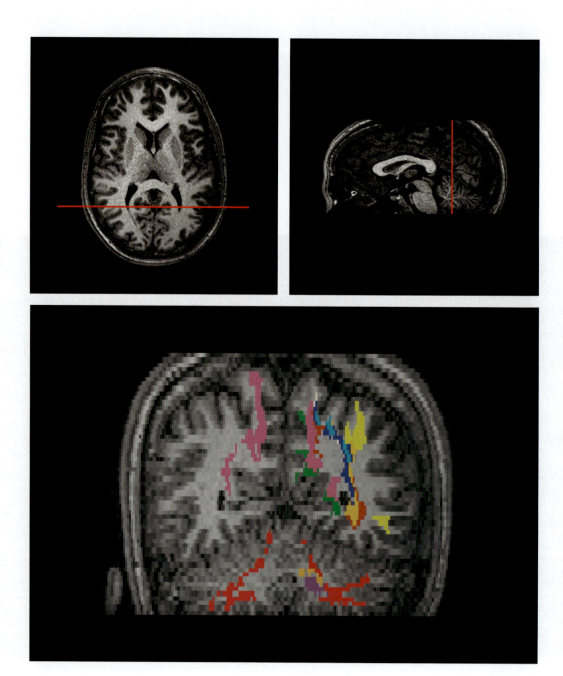

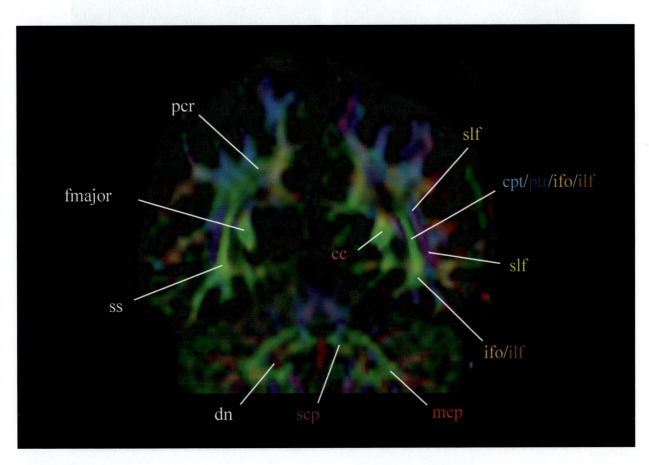

cc: corpus callosum
cpt: corticopontine tract
dn: dentate nucleus
fmajor: forceps major
ifo: inferior fronto-occipital fasciculus
ilf: inferior longitudinal faciculus

mcp: middle cerebellar peduncle
pcr: posterior corona radiata
ptr: posterior thalamic radiation
scp: superior cerebellar peduncle
slf: superior longitudinal fasciculus
ss: sagittal stratum

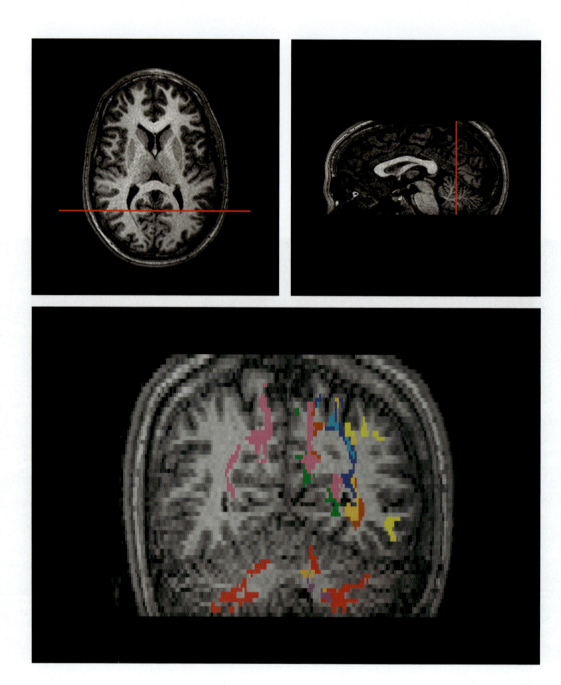

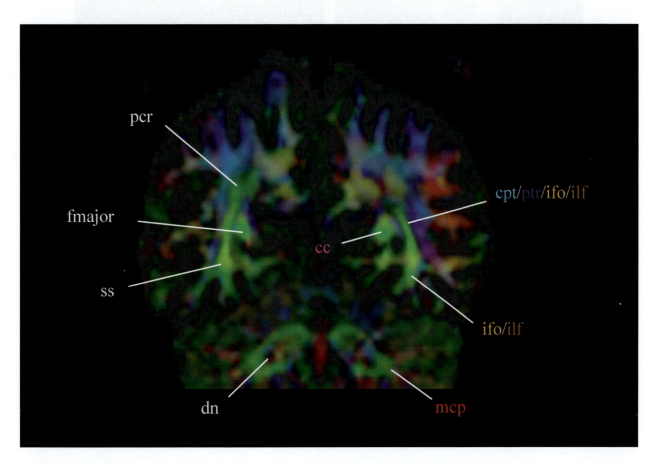

cc: corpus callosum
cpt: corticopontine tract
dn: dentate nucleus
fmajor: forceps major
ifo: inferior fronto-occipital fasciculus

ilf: inferior longitudinal faciculus
mcp: middle cerebellar peduncle
pcr: posterior corona radiata
ptr: posterior thalamic radiation
ss: sagittal stratum

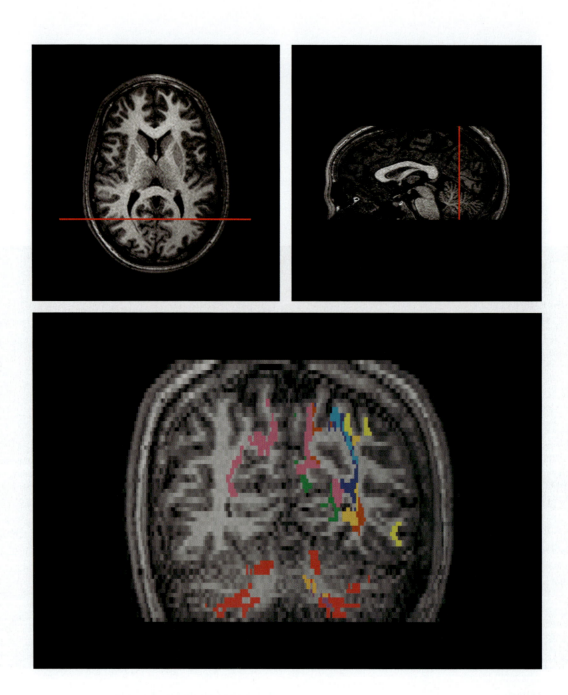

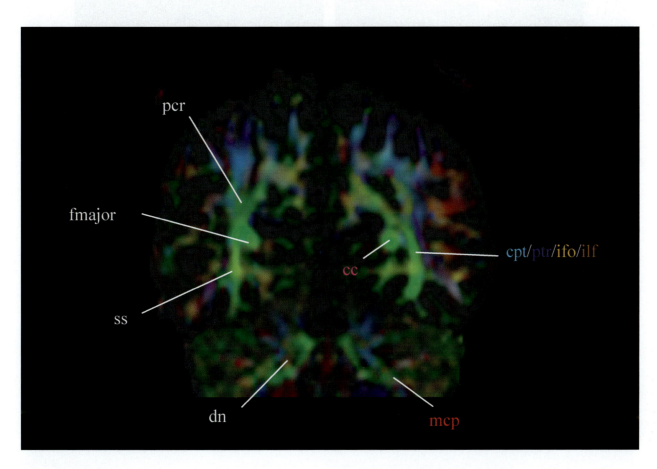

cc: corpus callosum
cpt: corticopontine tract
dn: dentate nucleus
fmajor: forceps major
ifo: inferior fronto-occipital fasciculus

ilf: inferior longitudinal faciculus
mcp: middle cerebellar peduncle
pcr: posterior corona radiata
ptr: posterior thalamic radiation
ss: sagittal stratum

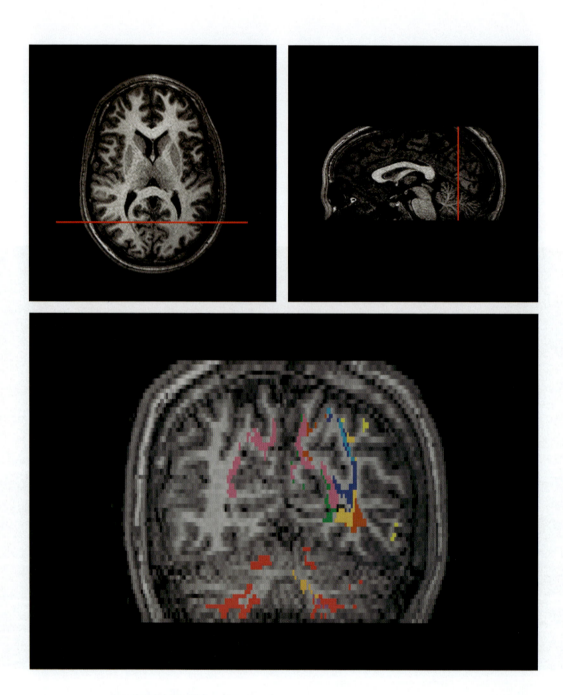

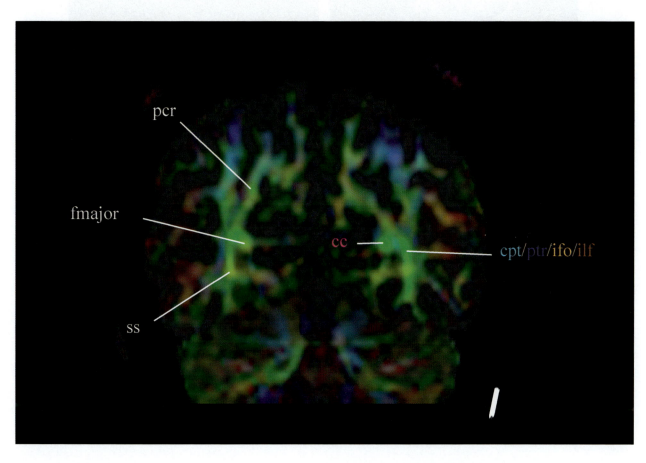

cc: corpus callosum
cpt: corticopontine tract
fmajor: forceps major
ifo: inferior fronto-occipital fasciculus

ilf: inferior longitudinal faciculus
pcr: posterior corona radiata
ptr: posterior thalamic radiation
ss: sagittal stratum

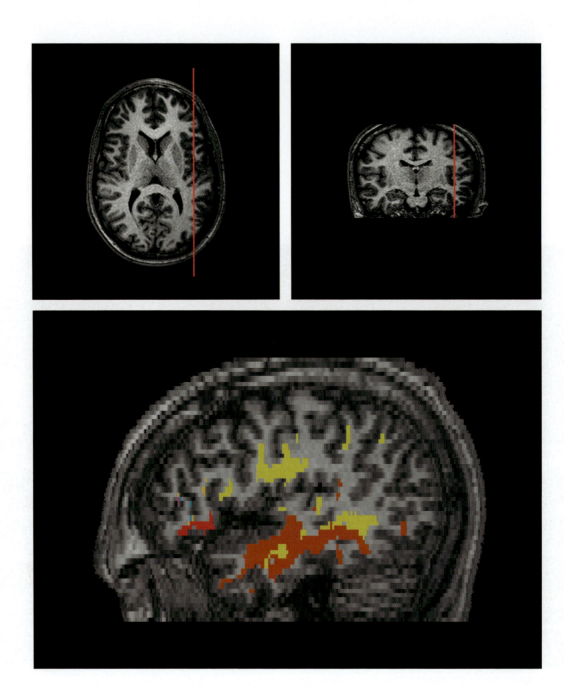

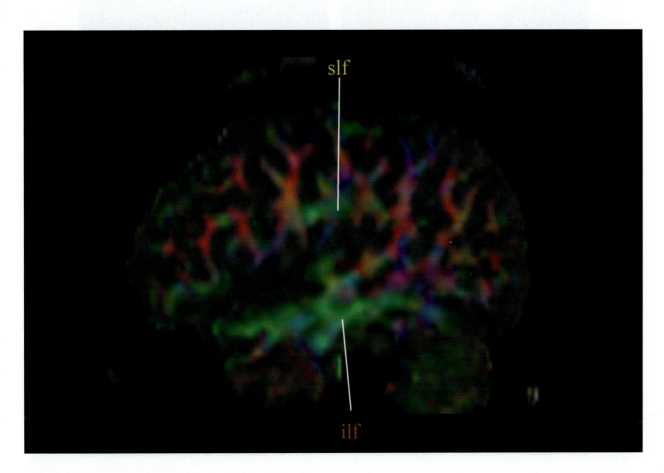

ilf: inferior longitudinal fasciculus
slf: superior longitudinal fasciculus

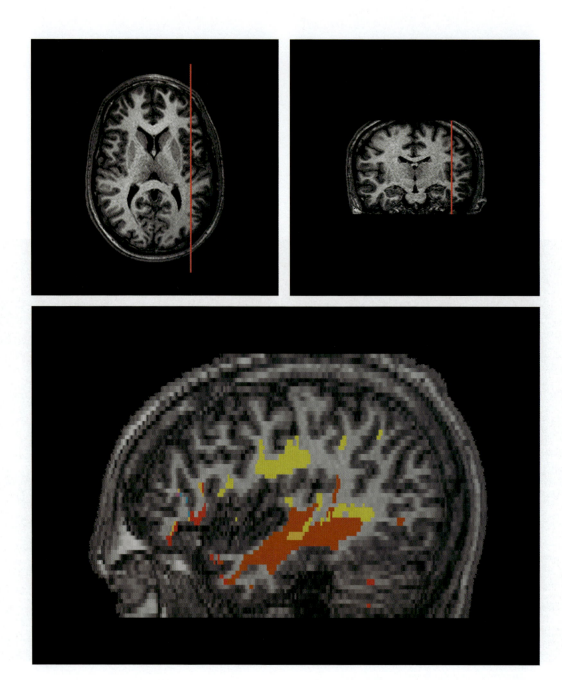

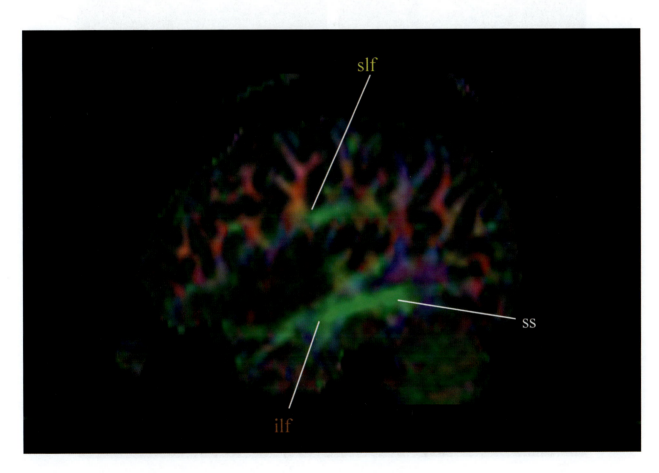

ilf: inferior longitudinal fasciculus   ss: sagittal stratum
slf: superior longitudinal fasciculus

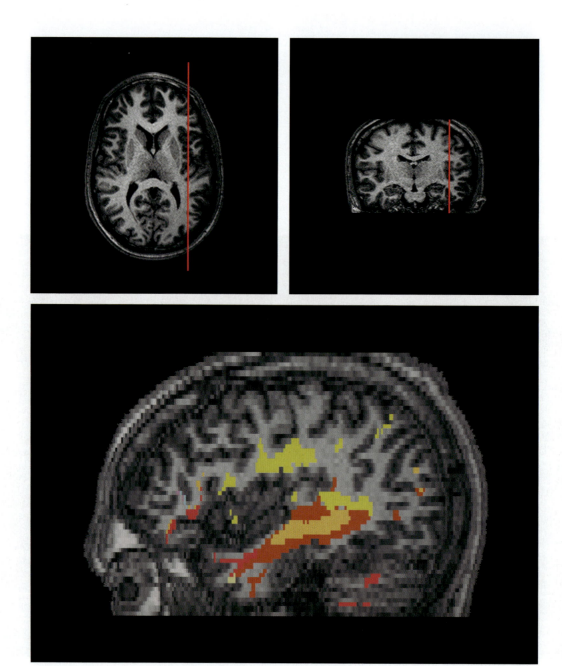

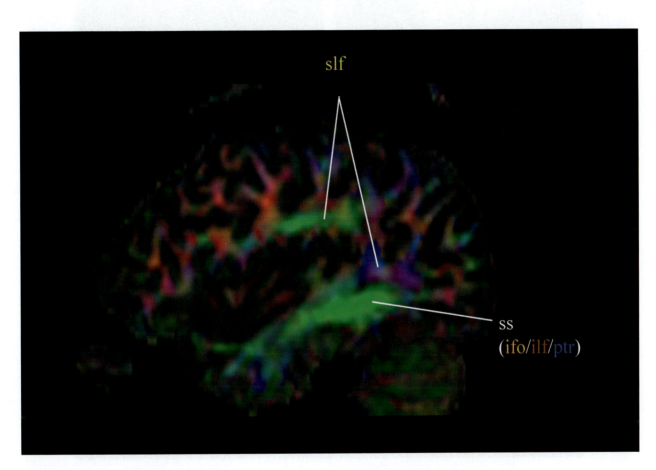

ifo: inferior fronto-occipital fasciculus
ilf: inferior longitudinal fasciculus
ptr: posterior thalamic radiation

slf: superior longitudinal fasciculus
ss: sagittal stratum

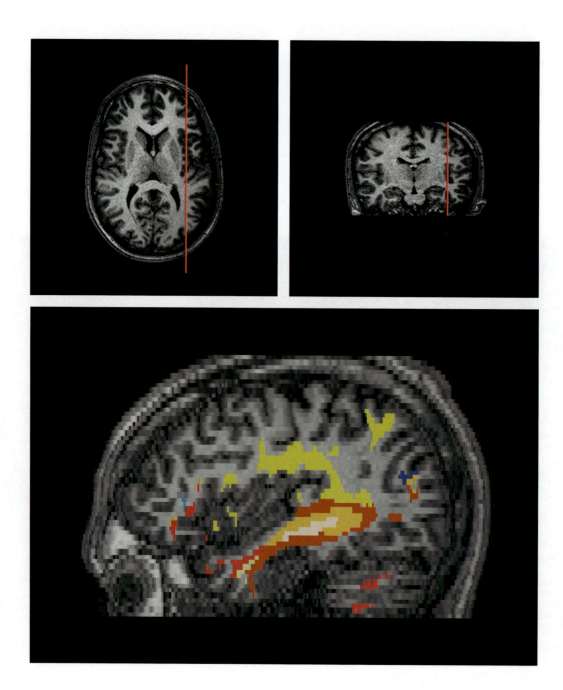

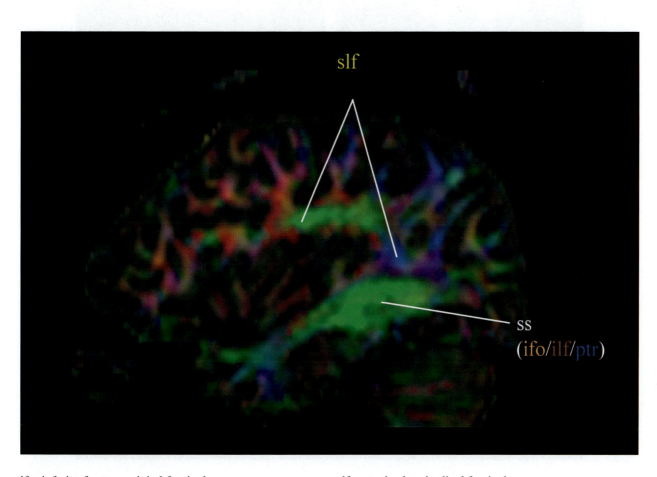

ifo: inferior fronto-occipital fasciculus
ilf: inferior longitudinal fasciculus
ptr: posterior thalamic radiation

slf: superior longitudinal fasciculus
ss: sagittal stratum

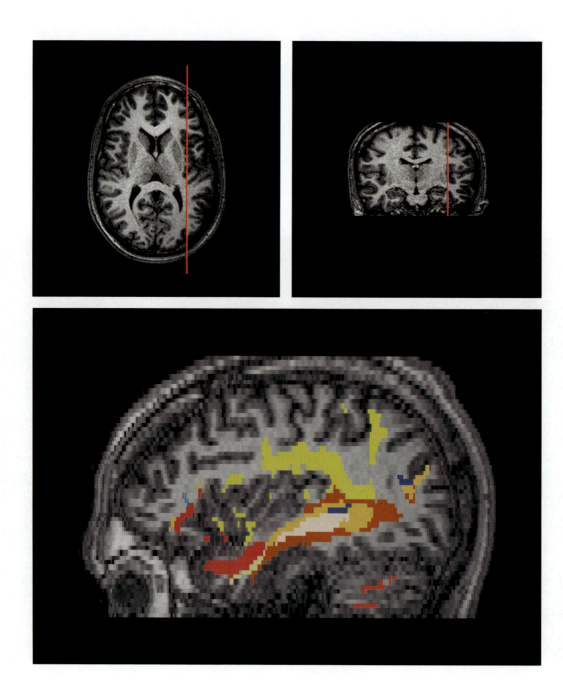

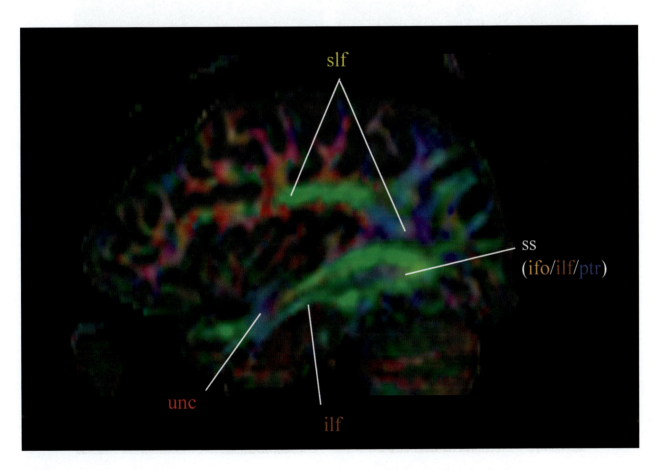

ifo: inferior fronto-occipital fasciculus   slf: superior longitudinal fasciculus
ilf: inferior longitudinal fasciculus   ss: sagittal stratum
ptr: posterior thalamic radiation   unc: uncinate fasciculus

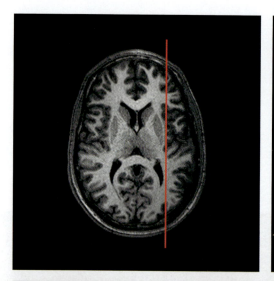
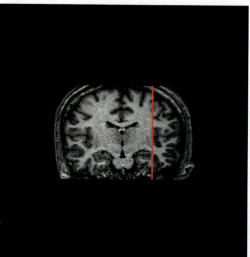
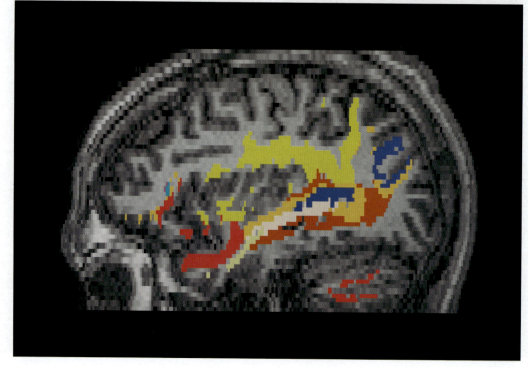

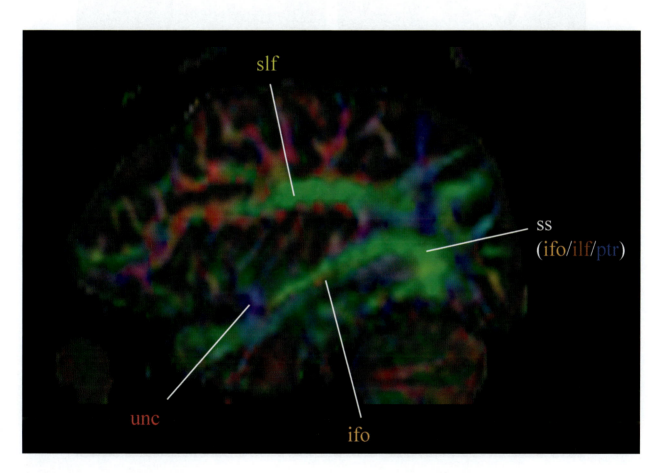

ifo: inferior fronto-occipital fasciculus
ilf: inferior longitudinal fasciculus
ptr: posterior thalamic radiation

slf: superior longitudinal fasciculus
ss: sagittal stratum
unc: uncinate fasciculus

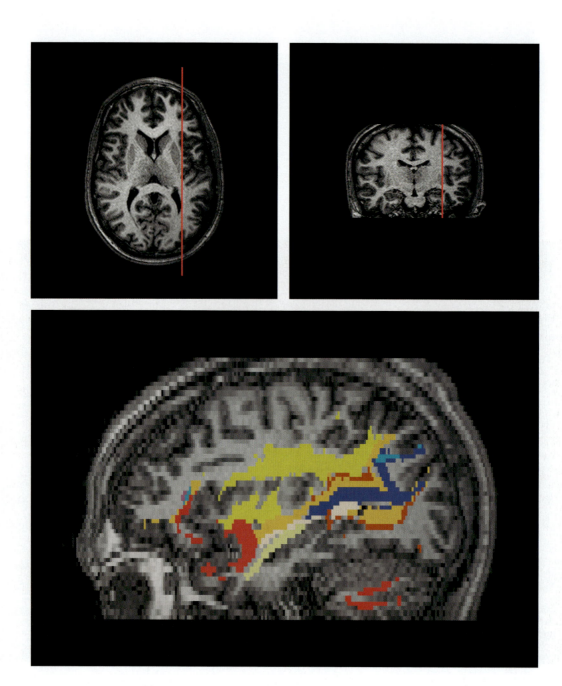

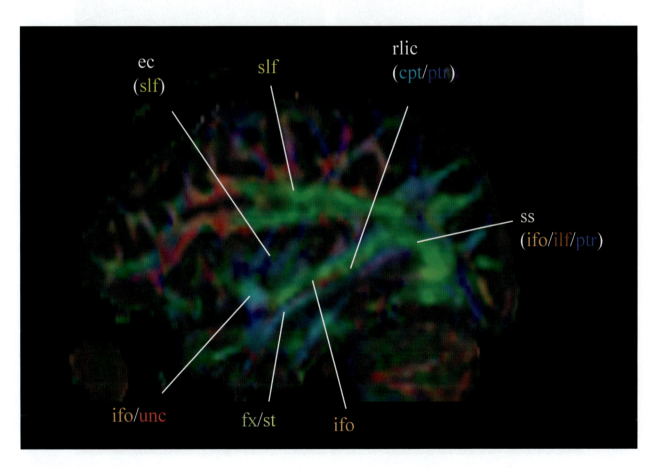

cpt: corticopontine tract
ec: external capsule
fx: fornix
ifo: inferior fronto-occipital fasciculus
ilf: inferior longitudinal fasciculus
ptr: posterior thalamic radiation

rlic: retrolenticular part of the internal capsule
slf: superior longitudinal fasciculus
ss: sagittal stratum
st: stria terminalis
unc: uncinate fasciculus

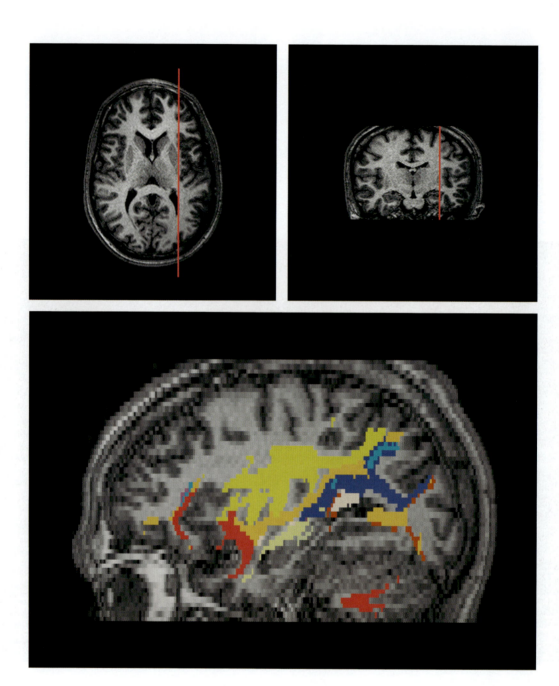

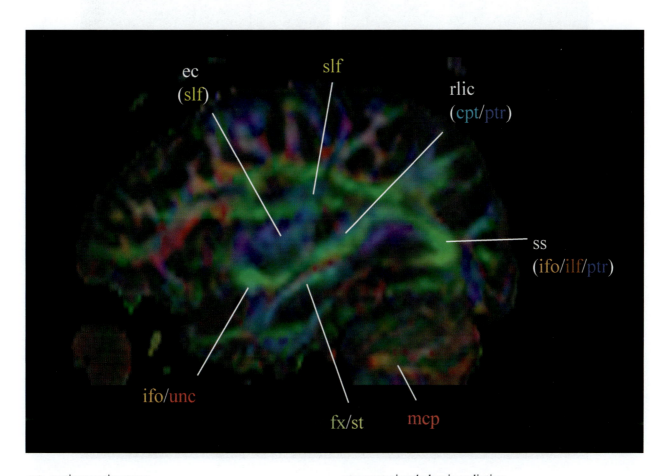

cpt: corticopontine tract
ec: external capsule
fx: fornix
ifo: inferior fronto-occipital fasciculus
ilf: inferior longitudinal fasciculus
mcp: middle cerebellar peduncle

ptr: posterior thalamic radiation
rlic: retrolenticular part of the internal capsule
slf: superior longitudinal fasciculus
ss: sagittal stratum
st: stria terminalis
unc: uncinate fasciculus

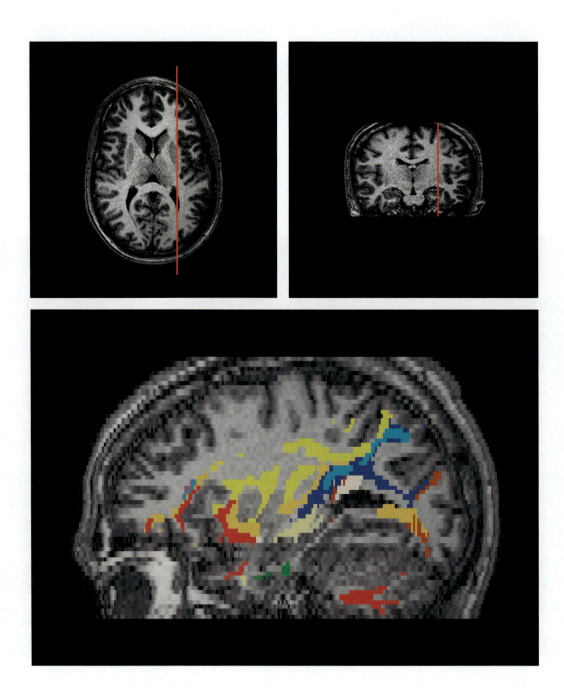

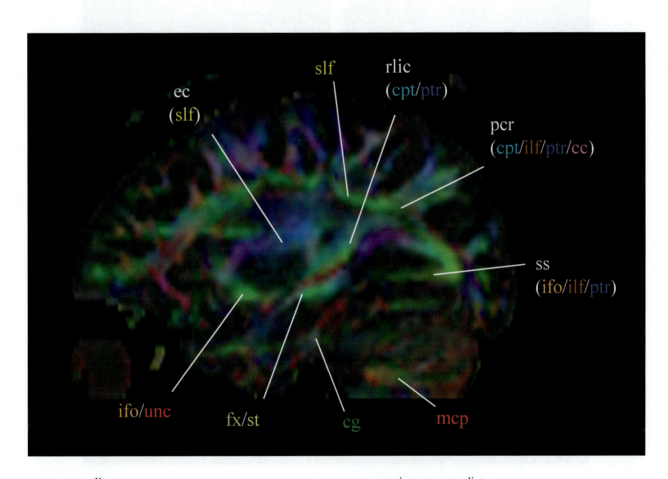

cc: corpus callosum
cg: cingulum
cpt: corticopontine tract
ec: external capsule
fx: fornix
ifo: inferior fronto-occipital fasciculus
ilf: inferior longitudinal fasciculus
mcp: middle cerebellar peduncle

pcr: posterior corona radiata
ptr: posterior thalamic radiation
rlic: retrolenticular part of the internal capsule
slf: superior longitudinal fasciculus
ss: sagittal stratum
st: stria terminalis
unc: uncinate fasciculus

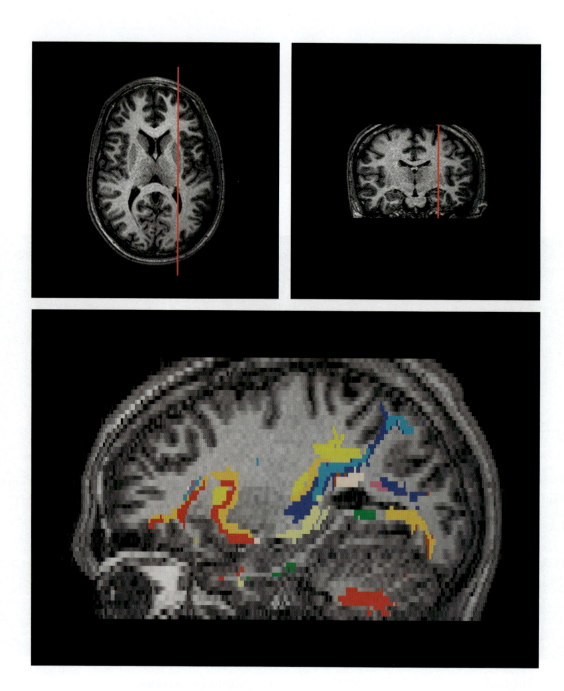

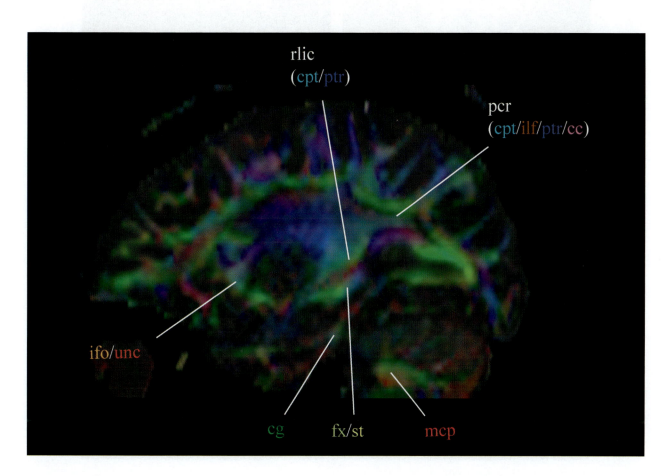

cc: corpus callosum
cg: cingulum
cpt: corticopontine tract
fx: fornix
ifo: inferior fronto-occipital fasciculus
ilf: inferior longitudinal fasciculus

mcp: middle cerebellar peduncle
pcr: posterior corona radiata
ptr: posterior thalamic radiation
rlic: retrolenticular part of the internal capsule
st: stria terminalis
unc: uncinate fasciculus

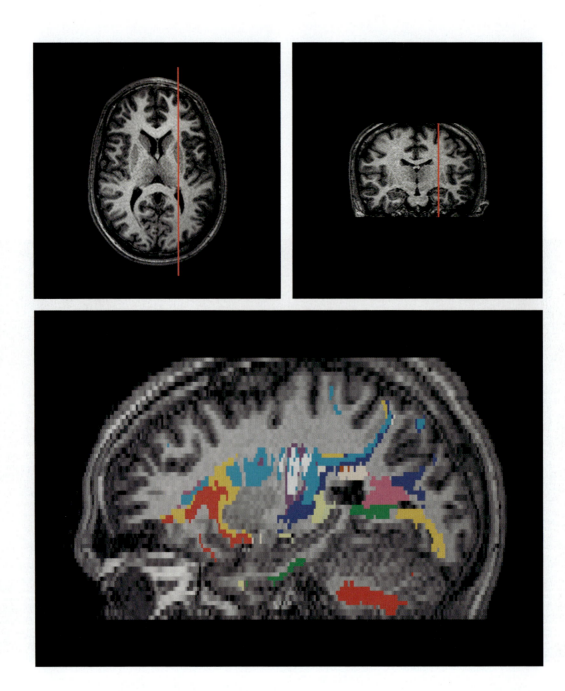

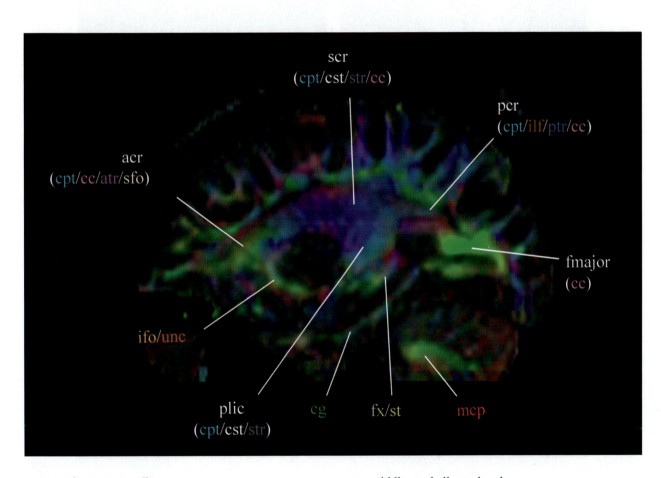

acr: anterior corona radiata
atr: anterior thalamic radiation
cc: corpus callosum
cg: cingulum
cpt: corticopontine tract
cst: corticospinal tract
fmajor: forceps major
fx: fornix
ifo: inferior fronto-occipital fasciculus
ilf: inferior longitudinal fasciculus

mcp: middle cerebellar peduncle
pcr: posterior corona radiata
plic: posterior limb of the internal capsule
ptr: posterior thalamic radiation
scr: superior corona radiata
sfo: superior fronto-occipital fasciculus
st: stria terminalis
str: superior thalamic radiation
unc: uncinate fasciculus

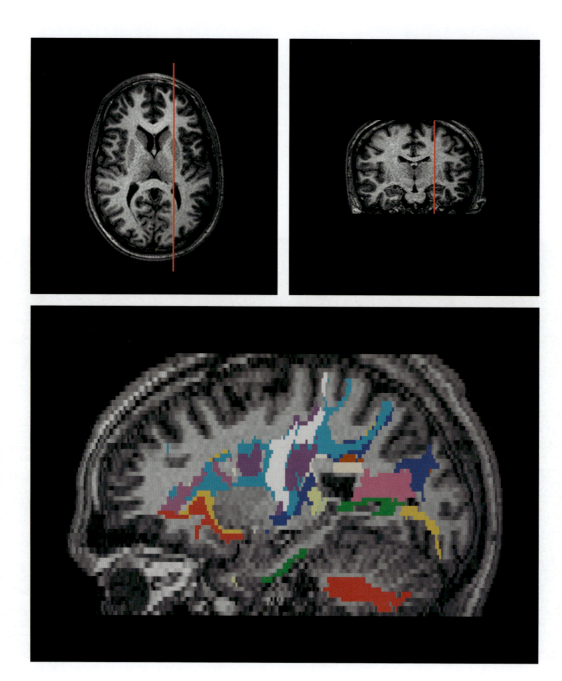

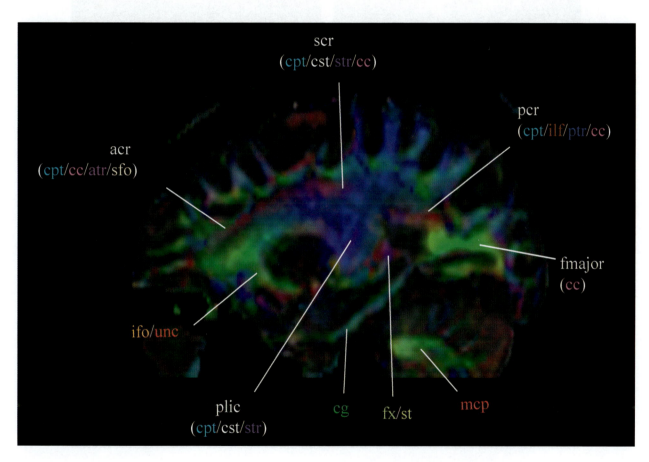

acr: anterior corona radiata
atr: anterior thalamic radiation
cc: corpus callosum
cg: cingulum
cpt: corticopontine tract
cst: corticospinal tract
fmajor: forceps major
fx: fornix
ifo: inferior fronto-occipital fasciculus
ilf: inferior longitudinal fasciculus

mcp: middle cerebellar peduncle
pcr: posterior corona radiata
plic: posterior limb of the internal capsule
ptr: posterior thalamic radiation
scr: superior corona radiata
sfo: superior fronto-occipital fasciculus
st: stria terminalis
str: superior thalamic radiation
unc: uncinate fasciculus

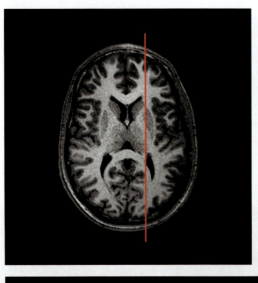
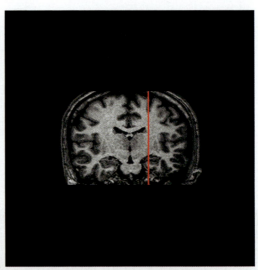
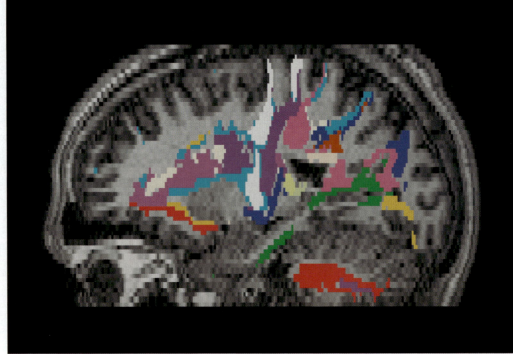

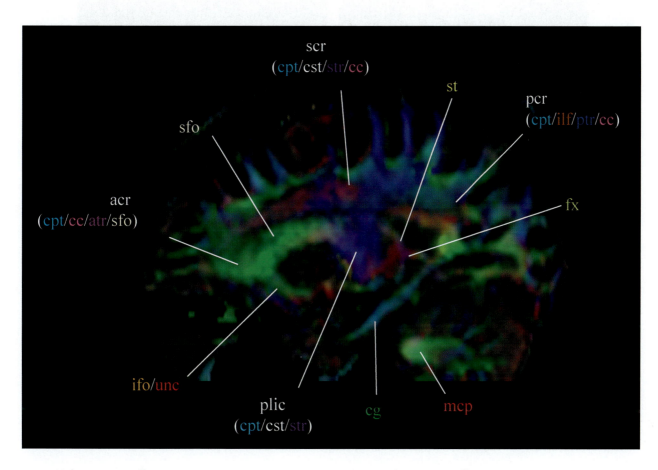

acr: anterior corona radiata
atr: anterior thalamic radiation
cc: corpus callosum
cg: cingulum
cpt: corticopontine tract
cst: corticospinal tract
fx: fornix
ifo: inferior fronto-occipital fasciculus
ilf: inferior longitudinal fasciculus
mcp: middle cerebellar peduncle

pcr: posterior corona radiata
plic: posterior limb of the internal capsule
ptr: posterior thalamic radiation
scr: superior corona radiata
sfo: superior fronto-occipital fasciculus
st: stria terminalis
str: superior thalamic radiation
unc: uncinate fasciculus

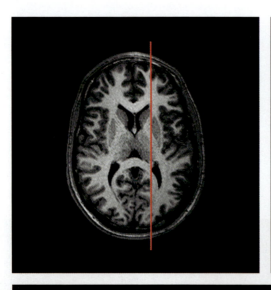
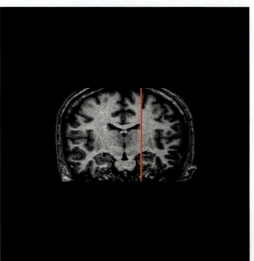
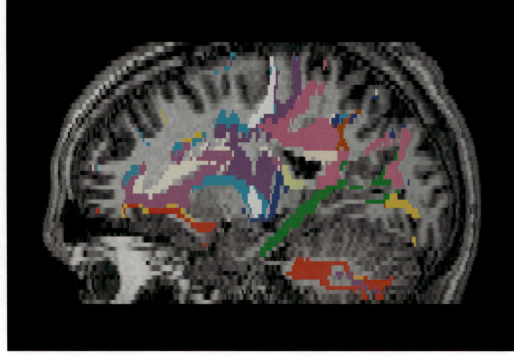

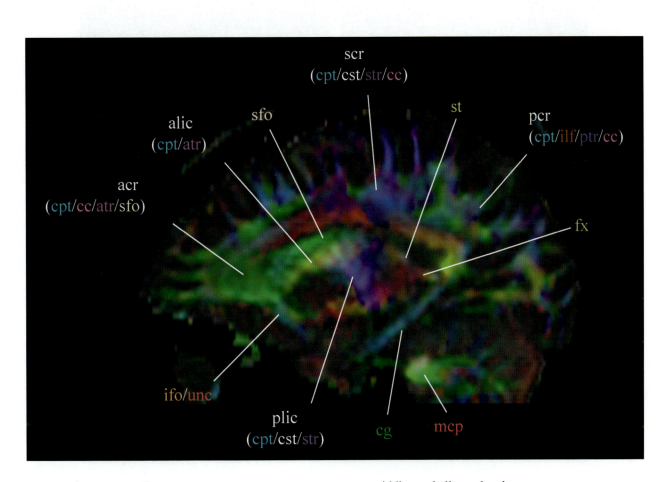

acr: anterior corona radiata
alic: anterior limb of the internal capsule
atr: anterior thalamic radiation
cc: corpus callosum
cg: cingulum
cpt: corticopontine tract
cst: corticospinal tract
fx: fornix
ifo: inferior fronto-occipital fasciculus
ilf: inferior longitudinal fasciculus

mcp: middle cerebellar peduncle
pcr: posterior corona radiata
plic: posterior limb of the internal capsule
ptr: posterior thalamic radiation
scr: superior corona radiata
sfo: superior fronto-occipital fasciculus
st: stria terminalis
str: superior thalamic radiation
unc: uncinate fasciculus

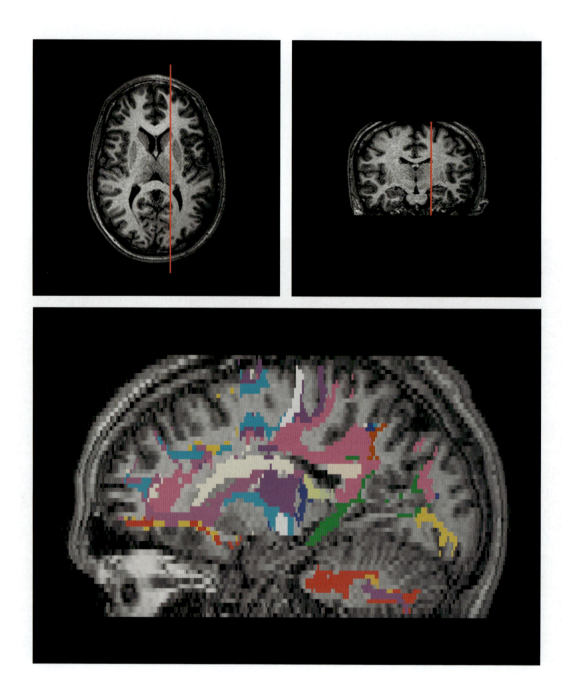

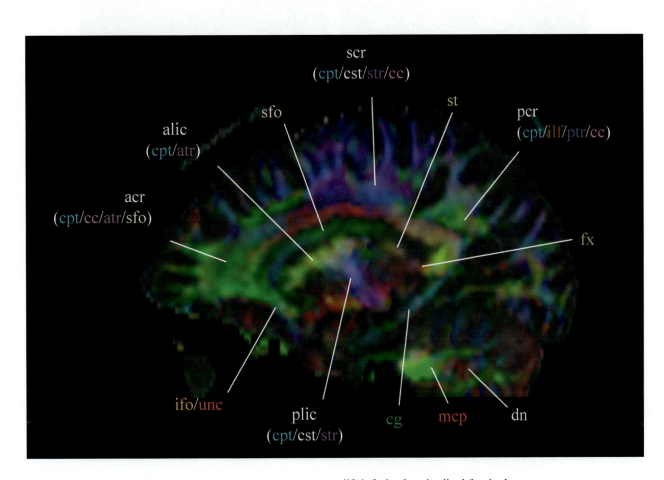

acr: anterior corona radiata
alic: anterior limb of the internal capsule
atr: anterior thalamic radiation
cc: corpus callosum
cg: cingulum
cpt: corticopontine tract
cst: corticospinal tract
dn: dentate nucleus
fx: fornix
ifo: inferior fronto-occipital fasciculus

ilf: inferior longitudinal fasciculus
mcp: middle cerebellar peduncle
pcr: posterior corona radiata
plic: posterior limb of the internal capsule
ptr: posterior thalamic radiation
scr: superior corona radiata
sfo: superior fronto-occipital fasciculus
st: stria terminalis
str: superior thalamic radiation
unc: uncinate fasciculus

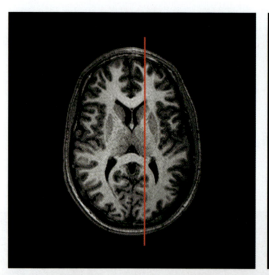
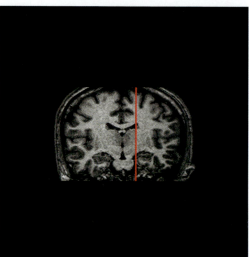
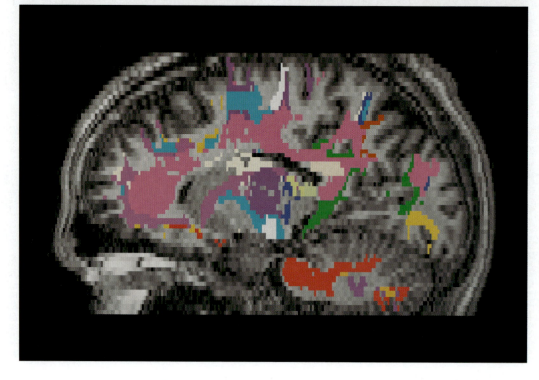

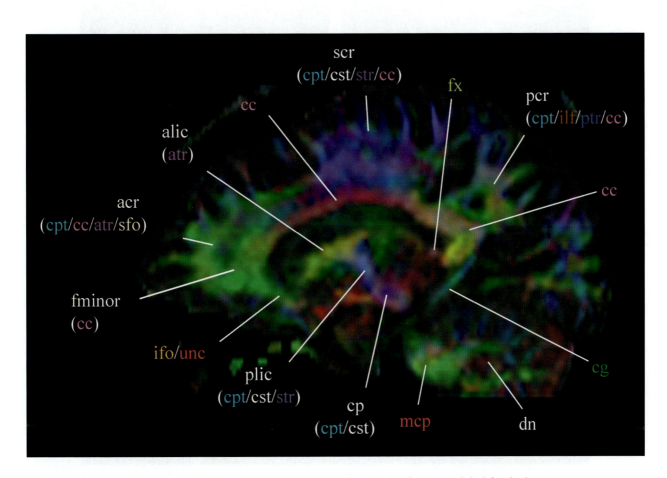

acr: anterior corona radiata
alic: anterior limb of the internal capsule
atr: anterior thalamic radiation
cc: corpus callosum
cg: cingulum
cp: cerebral peduncle
cpt: corticopontine tract
cst: corticospinal tract
dn: dentate nucleus
fminor: forceps minor
fx: fornix

ifo: inferior fronto-occipital fasciculus
ilf: inferior longitudinal fasciculus
mcp: middle cerebellar peduncle
pcr: posterior corona radiata
plic: posterior limb of the internal capsule
ptr: posterior thalamic radiation
scr: superior corona radiata
sfo: superior fronto-occipital fasciculus
str: superior thalamic radiation
unc: uncinate fasciculus

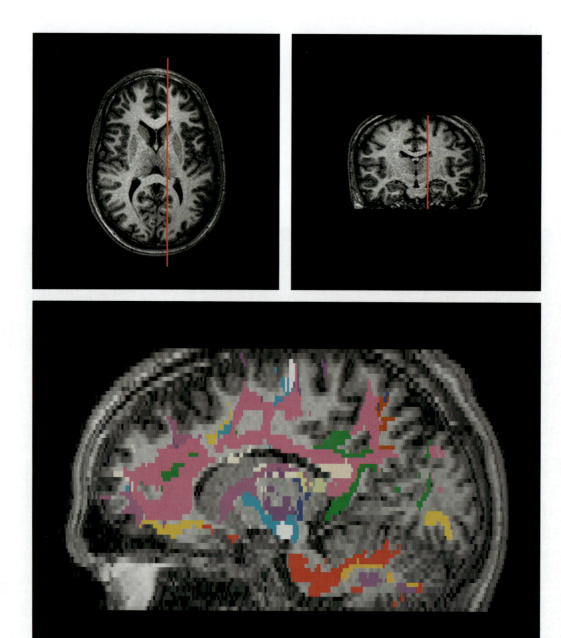

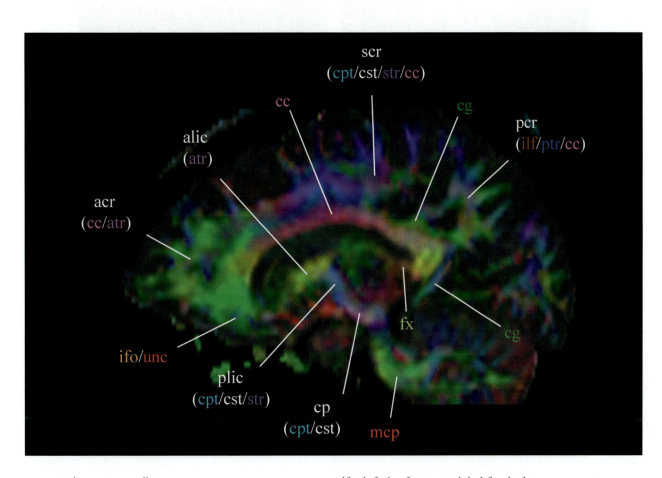

acr: anterior corona radiata
alic: anterior limb of the internal capsule
atr: anterior thalamic radiation
cc: corpus callosum
cg: cingulum
cp: cerebral peduncle
cpt: corticopontine tract
cst: corticospinal tract
fx: fornix

ifo: inferior fronto-occipital fasciculus
ilf: inferior longitudinal fasciculus
mcp: middle cerebellar peduncle
pcr: posterior corona radiata
plic: posterior limb of the internal capsule
ptr: posterior thalamic radiation
scr: superior corona radiata
str: superior thalamic radiation
unc: uncinate fasciculus

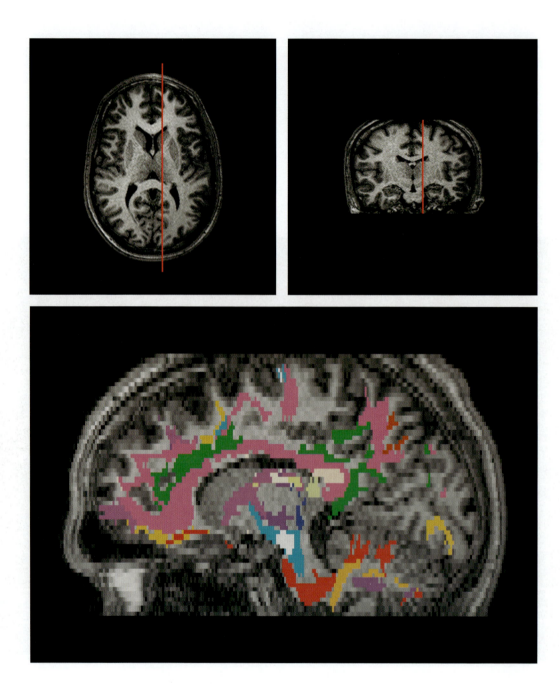

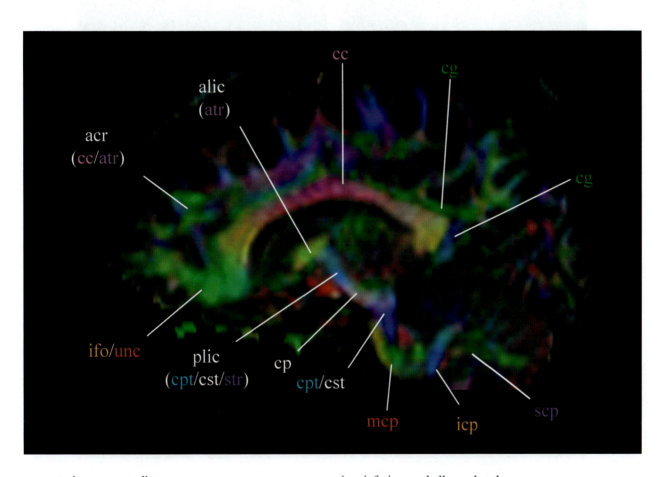

acr: anterior corona radiata
alic: anterior limb of the internal capsule
atr: anterior thalamic radiation
cc: corpus callosum
cg: cingulum
cp: cerebral peduncle
cpt: corticopontine tract
cst: corticospinal tract

icp: inferior cerebellar peduncle
ifo: inferior fronto-occipital fasciculus
mcp: middle cerebellar peduncle
plic: posterior limb of the internal capsule
scp: superior cerebellar peduncle
str: superior thalamic radiation
unc: uncinate fasciculus

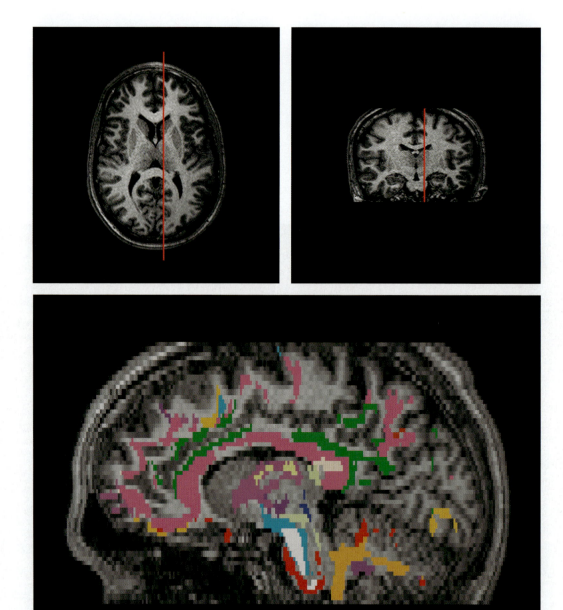

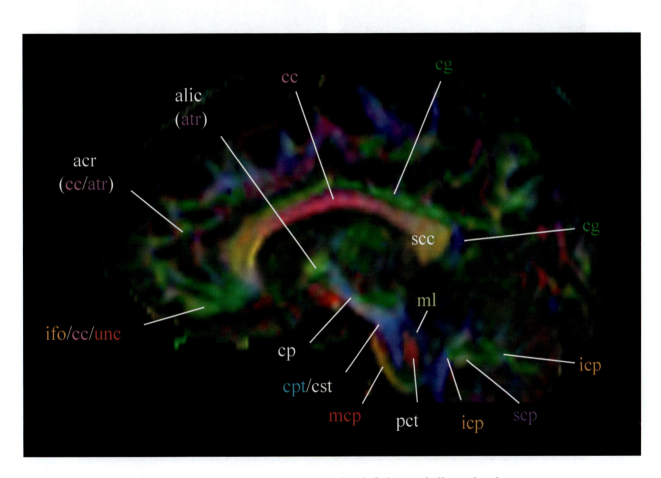

acr: anterior corona radiata
alic: anterior limb of the internal capsule
atr: anterior thalamic radiation
cc: corpus callosum
cg: cingulum
cp: cerebral peduncle
cpt: corticopontine tract
cst: corticospinal tract

icp: inferior cerebellar peduncle
ifo: inferior fronto-occipital fasciculus
mcp: middle cerebellar peduncle
ml: medial lemniscus
pct: pontine crossing tract
scc: splenium of the corpus callosum
scp: superior cerebellar peduncle
unc: uncinate fasciculus

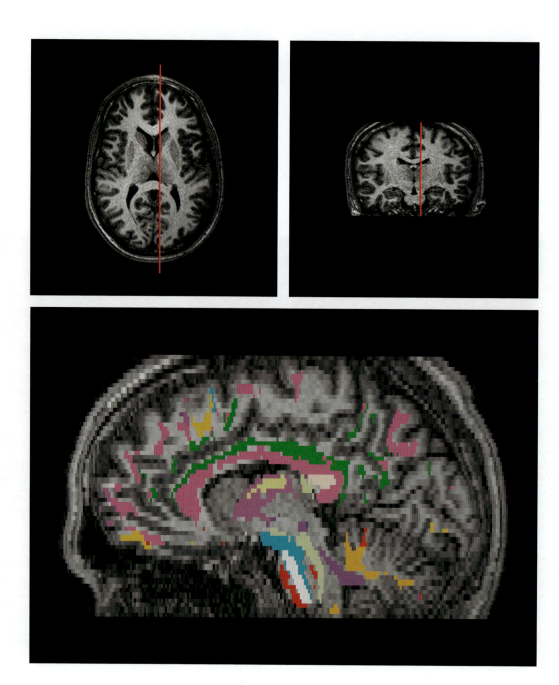

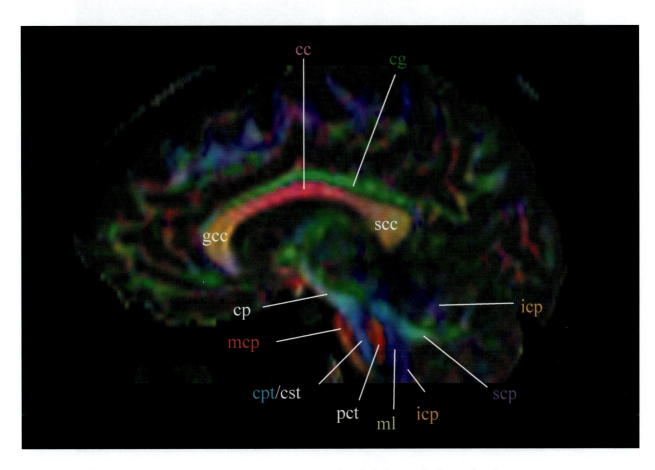

cc: corpus callosum
cg: cingulum
cp: cerebral peduncle
cpt: corticopontine tract
cst: corticospinal tract
gcc: genu of the corpus callosum

icp: inferior cerebellar peduncle
mcp: middle cerebellar peduncle
ml: medial lemniscus
pct: pontine crossing tract
scc: splenium of the corpus callosum
scp: superior cerebellar peduncle

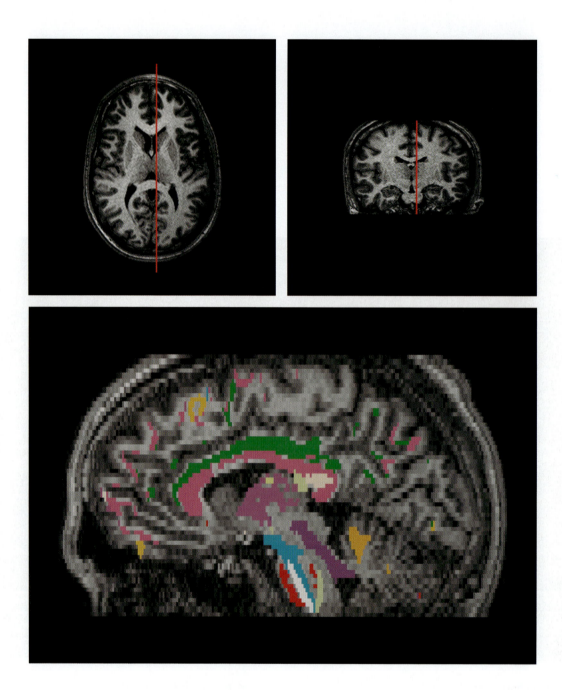

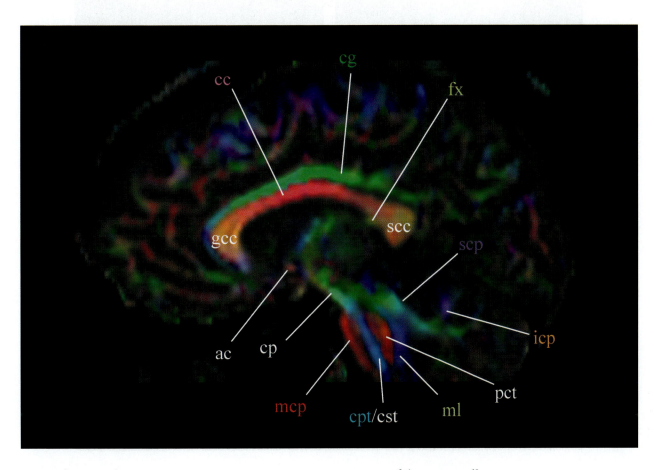

ac: anterior commissure
cc: corpus callosum
cg: cingulum
cp: cerebral peduncle
cpt: corticopontine tract
cst: corticospinal tract
fx: fornix

gcc: genu of the corpus callosum
icp: inferior cerebellar peduncle
mcp: middle cerebellar peduncle
ml: medial lemniscus
pct: pontine crossing tract
scc: splenium of the corpus callosum
scp: superior cerebellar peduncle

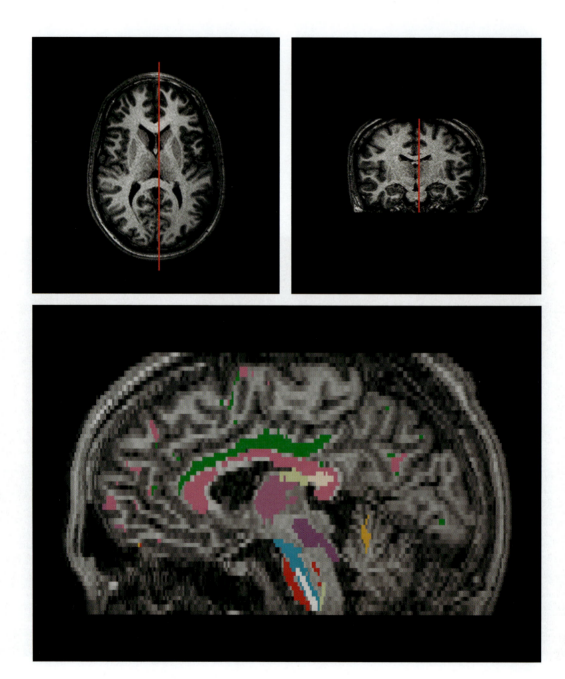

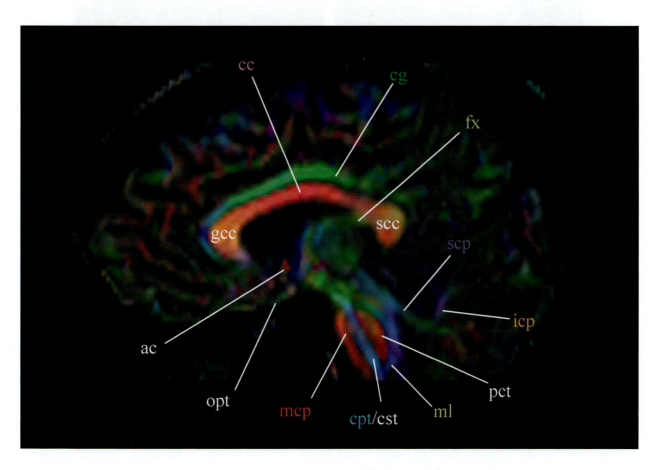

ac: anterior commissure
cc: corpus callosum
cg: cingulum
cpt: corticopontine tract
cst: corticospinal tract
fx: fornix
gcc: genu of the corpus callosum

icp: inferior cerebellar peduncle
mcp: middle cerebellar peduncle
ml: medial lemniscus
opt: optic tract
pct: pontine crossing tract
scc: splenium of the corpus callosum
scp: superior cerebellar peduncle

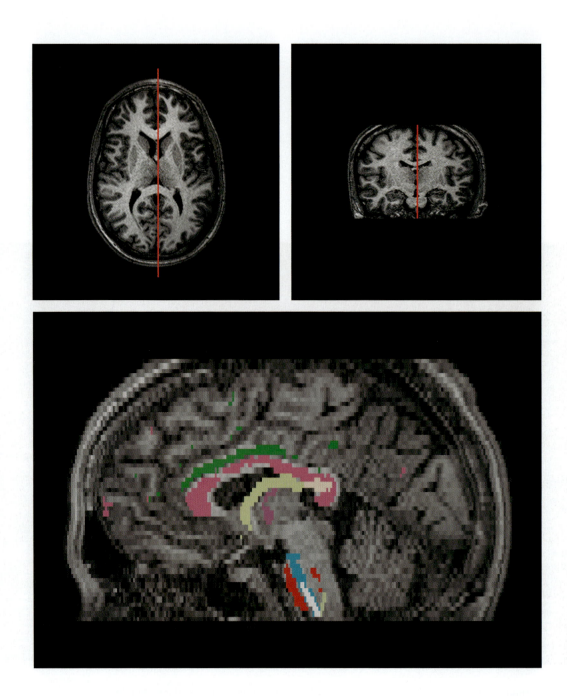

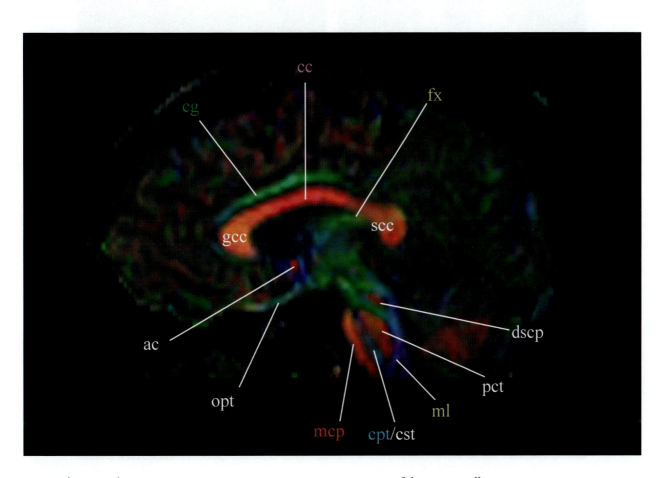

ac: anterior commissure

cc: corpus callosum

cg: cingulum

cpt: corticopontine tract

cst: corticospinal tract

dscp: decussation of the superior cerebellar peduncles

fx: fornix

gcc: genu of the corpus callosum

mcp: middle cerebellar peduncle

ml: medial lemniscus

opt: optic tract

pct: pontine crossing tract

scc: splenium of the corpus callosum

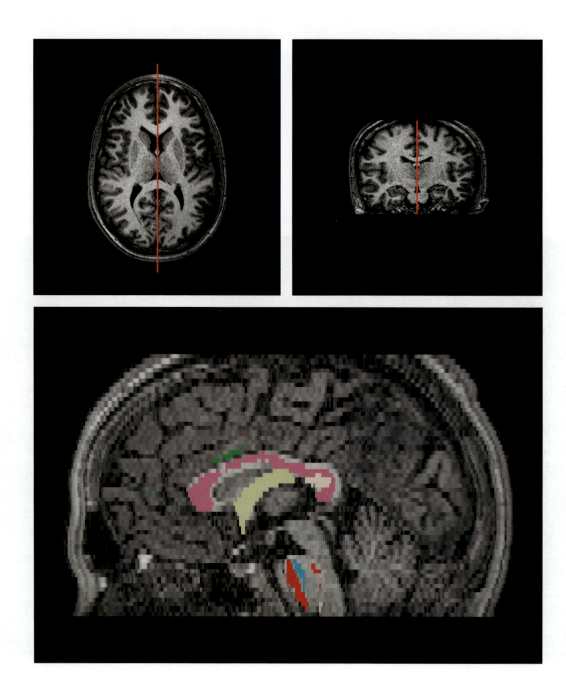

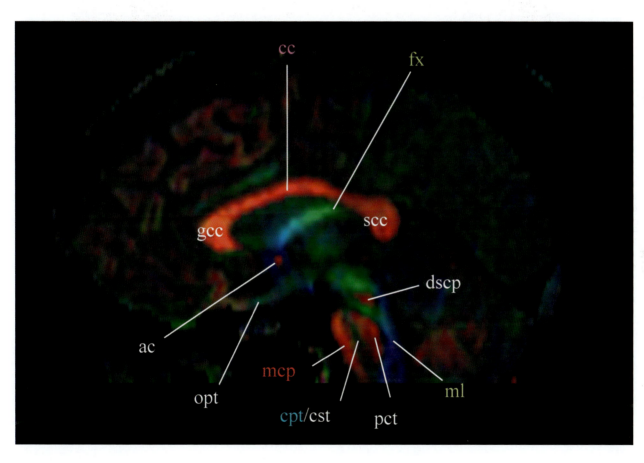

ac: anterior commissure
cc: corpus callosum
cpt: corticopontine tract
cst: corticospinal tract
dscp: decussation of the superior cerebellar peduncles
fx: fornix

gcc: genu of the corpus callosum
mcp: middle cerebellar peduncle
ml: medial lemniscus
opt: optic tract
pct: pontine crossing tract
scc: splenium of the corpus callosum

Subject Index